THEY BUILT AN INDUSTRY

The History of the Pressure Sensitive Adhesive Roll Label Industry

BY

BILL KLEIN

And 125 Members of the Industry

Development of this book was funded, in part, through the generosity of the Tag and Label Manufacturers Institute, Inc.

WITH

BRUCE AND VIRGINIA RONALD

FAIRWAY PRESS
LIMA, OHIO

THEY BUILT AN INDUSTRY

FIRST EDITION

ISBN 1-55673-956-7

PRINTED IN U.S.A.

To my wife, Polly, a true helpmate in all my ventures, including "They Built An Industry."

ACKNOWLEDGMENTS

I owe a debt of gratitude to Don McDaniel of MPI for convincing me to write "They Built An Industry." It began as a chore and ended up as a labor of love. I am also grateful to all of the men and women of the industry who gave of their time and experiences in participating in the writing of this book. You will find all of their names and stories within these pages.

A special thanks goes to Al Norman for his strong support, technical expertise and time spent in editing to correct the spelling of names of some of the early pioneers and to confirm the technical evolution of pressure sensitive adhesive coating and materials.

I am equally grateful to Duane Polkinghorne for his editing support and counsel throughout the process of writing the book and to Jim Kyte for his editing and insights into the industry's evolution and growth to greater sophistication and technical capability.

The viability of this project has been significantly enhanced by the contributions of:

MPI - initial funding and continuing support
TLMI - co-sponsorship funding and technical support
E.I. DuPont - (Susan Kremer) dust cover design
Simpson Paper Company - contribution of paper stocks:
Simpson Ever Green Gloss 70# (photo pages)
Simpson Ever Green Gloss (80#) (dust jacket)
Simpson Satin - Kote 60# (text pages)

And, finally, but not in order of significance, I wish to thank Bruce and Virginia Ronald for their significant writing, organizing and editing contributions to "They Built An Industry."

PREFACE BY STAN AVERY

Over the years, it has been a source of awe and amazement to me that the concept of a label that could stick without moistening would start an entire industry, one broad enough in scope and application to create a host of related products and businesses and provide countless people with jobs and livelihoods.

I'm often asked if I had a grand vision in which the technology of the pressure-sensitive label business would burgeon into a new industry. The fact is, I did not. My invention of the self-adhesive label in 1935 was simply the result of trying to find a way to solve a very specific problem. It was in the middle of the Big Depression and I was only semi-employed packing flowers at the San Lorenzo Nursery in Los Angeles. I was just fortunate enough to create a product that grew into a large, healthy business as more and more uses and applications for self-adhesive materials were discovered. And, in retrospect, I have to say that there is probably no better time to start a business than under those circumstances. When you have nothing to lose, you take risks without even thinking about it.

But it wasn't a smooth ride. I had my share of disappointments, bad times and obstacles to surmount. Money and time for making improvements were scarce. Nevertheless, in such spare time as I could find, I developed an entirely new and better method for making labels by die-cutting them in any desired shape right on the protective backing sheet rather than depositing them with a costly die-punch. It was a method which made it possible to manufacture the laminated material in bulk at high speed as an entirely separate operation. It is that method which is the foundation of the self-adhesive industry.

It was actually just good old "Yankee ingenuity," which runs in my family, that enabled me to take the idea and create this wonderful, profitable business and industry we have today. As time goes on, I am continually astounded at how the pressure-sensitive adhesive technology and products keep evolving. Today, our products decorate automobiles, label plastic containers, become self-adhesive stamps, and even take the place of paint on homes and buildings. As we constantly discover new and better applications for our technologies, the potential for self-adhesive products just grows and grows. I am extremely proud to have actually originated this ongoing, exciting and forever expanding industry.

Stan Avery
January, 1994

FOREWORD BY DON MCDANIEL

My association with pressure sensitive roll labels dates back to 1956 when I joined Salem Label as their first salesman. I had grown up on a farm near Sebring, Ohio, and began my business career selling livestock feed, then moving on to sell insurance for a while, when the opportunity came up to sell printing and labels for Salem Label.

At that time, pressure sensitive adhesive roll labels were just a side-line, they were almost considered as a novelty within the packaging and graphic arts field. When I began I knew nothing about the product, its usefulness or its potential. It was not very long before I began to develop new applications and find new customers. In fact, I found it easy and fun to sell pressure sensitive labels. So much so, that in 1967 I decided to buy a used Mark Andy press and go out on my own, concentrating exclusively on the production and sale of pressure sensitive roll labels.

Looking back, I admit it was a humble beginning. But, like so many others who have built their careers in this industry, I prospered beyond my wildest dreams. As I spoke with my suppliers and my associates in the industry I discovered they had the same appreciation for what happened in their careers because of pressure sensitive roll labels. This led to discussions of legacy. Not just what we could pass on to our families, which in many cases would be rather substantial, but what we could pass on to others who were just starting out. We wanted to tell the story of the building of this dynamic industry through the experiences and words of many of the people—the inventors, the dreamers, the tinkerers, farm boys, ex-GIs, chemists, engineers, accountants—the people who actually did the pioneering and building.

Finally, in mid-1990, I met Bill Klein, a consultant with extensive experience in the pressure sensitive industry. I convinced Bill to consider taking on our project to write the history of the pressure sensitive roll label industry. Bill spent the next twenty-five months taping the extensive interviews, researching and organizing the material to write **They Built An Industry**.

Don McDaniel
March, 1994

READER'S GUIDE

This book is the product of input from more than 125 members of the Pressure Sensitive Adhesive Roll Label (PSARL) industry. They range from the industry's early developers and leaders to the more recent shapers and molders of the industry. They were referred to Bill Klein by their peers. Unfortunately, a number of the people who were recommended to participate could not be located, *e.g.*, Dave Mages. After forty phone calls the search for him was abandoned. Of those who could be contacted, only three refused to participate. The rest agreed to be interviewed in depth by Bill Klein in person or over the telephone. A typewritten transcript of each interview was prepared, resulting in more than 3000 pages, or well over half a million words. Many other individuals are quoted as well, thanks to Avery's gracious permission to quote from their company history, The First Fifty Years. Material for that book was obtained in much the same way as this work.

But the important thing to note here is that with the exception of some extensive market research information on product application, technology and industry size, almost all the raw material of this book comes from the spoken memories of the people involved. Memories are fragile things and somewhat selective. What is important enough to be remembered can vary dramatically from individual to individual.

Most of the people in all phases of the PSARL industry were highly secretive, especially during the earlier years and, with the exception of those who worked in one firm and moved to another (or to start their own business), few had any direct contact with their competitors.

THE SOUNDING BOARD

Who is Bill Klein? He is currently a consultant working out of his farm near Springfield, Ohio. He worked for Fasson from 1960 to 1974 in Painesville, Ohio, and San Marino, California, and also in Tokyo, Japan as resident director of Sanyo/Fasson. He is a fervent believer in entrepreneurship in general and in the entrepreneurial nature of the PSARL industry in particular. His experience in the industry convinced many of the interviewees (some were downright testy at first) to open up and tell their stories as candidly as possible. People in this industry were secretive for so long that many were reluctant to name their early accounts or applications. Most, however, were surprisingly open with such information.

WARNING

Transcripts, especially transcripts from tapes recorded over the telephone, are tricky. The transcribers wrote what they heard but were not PSARL experts. Many industry figures and TLMI authorities have read the manuscript in order to correct as many errors and misspellings as possible. If no one caught it, we all apologize. A special thanks goes to Al Norman for his general support, encouragement and for reading the manuscript for technical accuracy related to materials and PSA coatings.

COMPLETELY ARBITRARY

Any book "written" by 125 sources is going to be fragmentary. To lessen this, the book is divided into sections chronologically:

1935-1945

1945-1955

1955-1960

1960-1970

1970-1980

1980-1990

1990-DATE

People, however, do not fit neatly into such a chronology. For the most part an interviewee is introduced and quoted the first time he (there are only two *she's*) entered the industry or started his own PSARL business. In some cases the interviewee is quoted in more than one section. Sometimes the interviewee is introduced and the timeline extended to present time. In a few instances the person may be introduced later during a time when he made—in our opinion—his most significant contribution to the industry.

All three of these techniques are arbitrary. They are an attempt to make a fragmentary manuscript less confusing. For example, had we quoted all those who cited the significance of photo polymer plates, we could have used quotes from at least forty people. We used only a few, but did mention the nearly universal recognition of this important advancement of plate technology.

All those interviewed for the project have been quoted, many extensively. Even where an indirect quote is used, the material surrounding the quote comes from information provided by the person interviewed.

You will occasionally find two people making identical claims to inventions. "We were the first to————" We quote them both. In the highly secretive atmosphere of the business, independent inventions or achievements were not unusual.

CHANGING TIMES

From the time the interviews were begun in July 1992 to the completion of a second draft in December 1993, the industry has undergone some dramatic changes:

MACtac acquired Fitchburg
Engraph is now a part of Sonoco
Rotometrics acquired Preston
Menasha acquired the Label Division of New Jersey Packaging

Only slightly earlier, Lord Label became a part of the English firm, Porter Chadburn. It may look like the industry is concentrating, yet the start-up opportunities are still extensive. For instance, according to equipment suppliers, both new and used, there are now dozens of die makers across the land instead of two or three, and forty to fifty new roll label converting companies starting up yearly! And, despite acquisitions on the material supply side, new starts and/or expansion by smaller coaters are currently underway to keep open the strongly competitive nature on the supply side of the industry.

A FINAL NOTE

How can you make your children become entrepreneurs? Give them two, three, four or five letter first names that can be pronounced with one syllable. How on earth did Percival Wise ever get into a business dominated by names like Stan, Mark, Burt, Chuck, Don, Al, Bill (never William), Dick (never Richard), Bob (never Robert), Lou, Ed, Joe, Walt, Carl, Jim, John, Tom, Rick, etc.?

The Pressure Sensitive Adhesive Roll Label industry is a fascinating example of entrepreneurship at its very best. Each day brings new challenges and opportunities. It is no wonder, that of all the people Bill Klein asked if they would do it all over again, not one said no.

Bruce and Virginia Ronald,
April, 1994.

The following are the 125 Contributors to the book, listed by the section in which they first appear

Anderson, Henry, 3
Andrews, Jr., Mark, 2
Andrews, Sr., Mark, 2
Avery, Stan, 1
Barnette, Joe, 5
Basto, Tricia, 5
Beaudoin, Leon, 4
Beck, Andy, 4
Bell, Bruce, 6
Benatar, Leo, 6
Buchta, Don, 1, 3
Bunnell, Dale, 2
Campbell, Alan, 4
Capuzzo, Dick, 2
Carlson, Lee, 2
Clifford, Paul, 5
Collons, George, 2
Corbett, George, 6
Daniell, Craig, 7
Deckard, Lon, 6
Desanzo, Dirk, 6
Dillingham, Bob, 6
Dochstader, Darrell, 5
Donovan, Bill, 3
Dunphy, Paul, 6
Eagon, Bev, 4
Egan, Dick, 3
Egbert, John, 6
Eiseman, Bill, 2
English, Jim, 1
Estridge, Bill, 4
Eyster, Don, 4
Farkas, Jules, 7
Ferris, Jim, 2
Finley, Bob, 4
Foukal, Don, 4
Fulweiler, Stan, 3
Garber, John, 2
Gartner, Gerry, 4
Good, David, 7
Gray, Bud, 4
Hailey, Lance, 6
Hart, Jim, 5
Hattemer, Jim, 5
Heintzelman, Skip, 7
Herrman, Mark, 5
Hillmer, Duane, 4
Hulton, Dennis, 5
Irwin, Bob, 2
Kennedy, Bert, 5
Kesten, Marty, 5
Kidd, Ken, 5
Klas, Bob, 2
Klein, Bill, 3
Klein, R.J., 6
Kunkle, Larry, 5
Künzle, Beda, 5
Kyte, Jim, 2
Lasinsky, Wayne, 2
Lewis, Gib, 4
Long, Robert P., 4
Louis, Maynard, 4
Mahnic, Ed, 2
Manzian, Peter, 5
Masterson, Patrick, 6
Mauritz, Dan, 5
McDaniel, Don, 3
McKillip, Barry, 7
McKinney, Gary, 6
Miller, Chuck, 7
Morgan, Burt, 3
Motter, Bruce, 4
Muir, Bill, 2
Nedblake, Wes, 7
Nerad, Jerry, 4
Noah, George, 5
Norman, Al, 3
Ordman, Les, 4
Orlando, John, 2
Parisi, George, 4
Patel, John, 6
Patrick, Pat, 6
Paul, Lee, 3
Pearson, Dick, 4
Percival, Alan, 6
Polkinghorne, Duane, 4
Prittie, Allan, 6
Questel, John M., 4
Reed, Chuck, 3
Reinke, George, 7
Riley, Dick, 3, 5
Rosemann, Richard, 4
Rüesch, Jr., Ferdinand, 5
Rüesch, Sr., Ferdinand, 2
Sauer, Jim, 6
Scheerer, Bob, 4
Scherer, Harold, 1
Schwartz, Art, 2
Schwartz, Dick, 6
Sidrane, Harold, 2
Singer, Gene, 4
Sirois, Bob, 7
Smith, Bob, 5
Smith, Herb, 2
Smith, Jim, 5
Sobel, Al, 2
Spear, Rick, 7
Spinello, Al, 5
Stillman, Harry, 2
Styers, Paul, 2
Swartz, Jack, 5
Tomlinson, Dan, 4
Trungale, Joe, 1
Vandenberg, Ed, 4
Weber, Joe, 6
Webster, Walt, 2
Werneke, Lou, 5
Wert, Mark, 4
Whitfield, Bill, 7
Williams, Carl, 5
Wilson, Jim, 3
Wilson, Tom, 3
Wise, Percival, 4
Yeagle, Paul, 6
Zimmer, Erik, 7
Zimmer, Karl, 7

TABLE OF CONTENTS

INTRODUCTION
Stan Avery v

FOREWORD
Don McDaniel vi

Reader's Guide vii
The Sounding Board vii
Warning viii
Completely Arbitrary viii
Changing Times ix
A Final Note x
List of Contributors xi

SECTION I: THE FIRST TEN YEARS: 1935-1945

The Label 3
One Man Industry 4
Depression 4
One-sided Gum Tacks 5
Angel in the Lab 6
New Plant and Fire 7
Heart of Flexographic 7
Innovations to 1944 9
Near Miss 9
Tangle with the Big Boys 10
The War 10
International Avery 11
Establishing the Mystique 12
Chronology of Applications, 1935-1945 13
Chronology of Inventions/Innovations, 1935-1945 13
The Man Who Wouldn't Quit, Part One 13

SECTION II: COMPETITION STIRS: 1945-1955

The Second Father	19
Christmas Tape	22
A Businessman Joins the Dreamer	22
The Three Markets	24
A Swiss Refusal	24
3M	25
Park Avenue Playboy	25
Shooting Labels	26
"Hello, My Name Is_________"	27
California Dreaming	28
The Whole Ball of Wax	29
Not Much Agony in Defeat	30
Coated Products	31
Competitors, Entrepreneurs, Innovators	33
No Small Potatoes	33
"Flood" Patents	34
"I Had to See How the Wheels Turned"	34
On His Own	36
Was It Profitable?	37
They Had Their Chance	37
Help Wanted	38
Prototype or Tinker Toy	39
Pressure Sensitive Adhesive Materials in the 1950s	40
A Rotary Person	41
Those Die Makers	42
Roto-Die	43
You Could Eat Your Mistakes	43
Entrepreneurs Wanted—$8.00 an Hour	45
Chronology of Applications, 1945-1955	46
Chronology of Inventions/Innovations 1945-1955	46
The Man Who Wouldn't Quit, Part Two	46

SECTION III: ALL THE PIECES IN PLACE: 1955-1960

The Giant Moves	53
International Avery	53
Fasson Becomes Avery's Competition	54
A Marx Brothers Movie	54

"We Destroyed Our Own Business" 57
Hazardous to Your Health 58
A Fire Truck A Day 58
The Price Is Right 59
-007- 60
Release Liners 61
Fasson Fantasia 61
Plain-Vu Ladies 62
Star Wars 63
The Ohio Year 64
All or Nothing 64
Poised for Growth 66
Grand Rapids 66
The Mysterious Debut of Label-Aire 67
Blades, Not the Razor 68
Here, There, Everywhere 68
The Smuggler 70
Go Where the Competition Isn't 71
Cherry Picking 72
And We Starved to Death 73
Easy to Remember Product Line 74
It Was Fun 76
Rah! Rah! Rah! 76
No Nice Young Men Available 77
Fifties Footnote 78
Chronology of Applications, 1955-1960 79
Chronology of Inventions/Innovations, 1955-1960 79
The Man Who Wouldn't Quit, Part Three 79

SECTION IV: LIFT OFF: 1960-1970

Explosive Growth 85
Infamous Third Generation 85
A Friend in Need 86
Liked the Location, the People, the Opportunity 86
Biggest Drivers 87
Thirty Year Man 88
Limited Automation 88
Irony of Ironies 89
Walk the Territory 90

The Last Hurrah 91
Marriage and Divorce 91
G or J? 92
Base Materials 93
Please Release Me 94
Exception to the Rule 95
You're Not Humble 95
Serendipity Time 96
Soabar, So Good 97
The Sock Fiasco 99
Do-It-Yourself 100
The User Friendly Press 101
You Owe Me 102
You're Unemployable 103
Dream Mind 105
Getting In On the Ground (Dirt) Floor 106
The Wizard of Ooze 107
From Juke Box to EDP 108
Niche Players 109
Just in Time 110
Still a Problem/Solution 111
Stick It To Me 112
Where's Your Lift Truck? 112
No Money, No Business, No Customers—Go For It 113
We Have a Week 115
One Man is the Production Line 116
Peel 'n Seal 116
"Hell" Riegels 118
I Like Competition 119
Computer Labels/Data Transfer 119
Applied Costs 120
Easy Money 120
Niche Players Form An Industry 121
Spring Housecleaning 122
Operator Driven 122
1. Buy Press 2. Start Company 122
Big Time 124
Slippery Chiquita 124
Stormin' Norman 125
Who You Gonna Call 126

No Moosic, Please 126
1000% Better 127
Star Over the Phone 128
A Year and a Half For Fitchburg 128
The Making of Millionaires 129
The Dynamic Dynamo 130
The Million Dollar Egg 131
Organizing the Industry 132
Tag Manufacturers Institute 132
The Roll Label Manufacturers Association 132
TLMI 133
Secrets 134
The Singer Report 135
Bob Klas on Gene Singer 136
Converter of the Year 137
LABELEXPO USA 137
International Cooperation 137
How Narrow is Narrow? 138
FTA 138
From Cannery Row to Modular Machines, Part I 139
Chronology of Applications, 1960-1970 140
Chronology of Inventions/Innovations, 1960-1970 141
The Man Who Wouldn't Quit, Part Four 141

SECTION V: QUALITY CREEPS IN: 1970-1980

Standardize, Standardize, Standardize 152
Swiss Connection 152
From Sub to Independent 154
Carloads of Two Mil Clear 155
Shoved Into the Marketplace 156
Under the Hood 157
More Base Material Improvements 158
Al Norman on Technical Breakthroughs in the 1970s 159
A Different Mind Set 160
Withering Heights 160
From Push to Pull 161
Better Things For Better Printing Through Chemistry 162
Obsolete on Arrival 163
Ink Spots 163

Learning Curve 165
And Then, Bar Codes 166
The Great Impetus 167
Instant Market Expansion 168
"Best Company I Ever Worked For" 168
It All Began in a Log Cabin 169
300 Miles More or Less 170
Philately to Pressure Sensitive 171
Almost a Disaster 172
A Family Affair 172
Credit to the People 174
Infinitely More Than Ink on Paper 175
Abe Reichman's Vice 177
Scared to Death 177
"I Wanted to Know More and More" 178
EPA and OSHA 179
From Zero to $22,000,000 in Six Years 180
A New Challenge Every Single Day 180
It Was Just Luck 181
Green Mountain Boys 181
Super Sharp 182
Consistency, Consistency, Consistency 182
The Company Bartering Built 184
In the Wee Small Hours of the Morning 185
Dark Secrets at PHL 185
Fifteen of Fifteen 186
Cottage for Sale 187
Four Hour Pay Off 187
That's Andy All Over 188
Unsung Heroines 188
Downsize to Survive 191
Leveling the Playing Field 192
The Teacher Who Didn't Want to Retire 193
You Just Can't Get By Anymore 194
Recession-Proof? 195
Avery International Corporation 196
The Seven Year Company 197
Bankruptcy Pays 199
Finding Intelligent Life 199

Chronology of Applications, 1970-1980 201
Chronology of Inventions/Innovations, 1970-1980 201
The Man Who Wouldn't Quit, Part Five 202

SECTION VI: COMPETITION IS THE HERO: 1980-1990

No Second Time 211
Constant Horror 212
Rather Be Lucky 212
Polite, But Still C.O.D. 214
Klein on Japan 215
There's Us and There's Avery 216
Again, A Recession Helps 216
Labeling System and Fly Rods 217
For Lasers Only 219
A Very Narrow Niche 220
The Forty Year Itch 221
Craftsmanship Versus Technology 222
Upstairs/Downstairs 223
An International War Room 224
Why 7 1/4 Inches? 224
Barely Enough to Buy Lunch 225
Thrashing Around, Barely Surviving 226
Turning Down Four Million 227
33 Employees, 300 Years' Experience 228
Watch Your Accounts Receivable 229
A New Giant 229
I Could Tell If I Had a Good Year By the Number of Cars in the Garage 231
Passing On the Torch 232
The Trouble Maker 232
It Would Take An Expert Among Experts 233
Unknown International Giant 233
No Longer Stand Alone Technology 234
A Prittie Press 234
Coaters 235
Laser Anilox Roll 235
A Slitter for Reykjavik 236
From Zero to $65,000,000 in Ten Years 236

Monarch For a Friend 237
A Plant and A Coater 238
Monarch 240
Dealing in a Time-Urgent Market 241
Entrepreneurs and Professional Managers 241
Super Bowl Trip Up 242
Bar Codes 243
From President to Lord 244
A Beautiful Marriage 245
Growing Pains 246
When You're Ready 247
Baby in Connecticut 247
Never Had So Much Fun in My Life 248
Exception to the Rule 248
Eight Presses, No Business 249
From Cannery Row to Modular Machines, Part II 250
U2 to Uarco 251
Not So Artsy-Craftsy 252
My Neighbor, the Buyer 252
Chronology of Applications, 1980-1990 253
Chronology of Inventions/Innovations, 1980-1990 253
The Man Who Wouldn't Quit, Part Six 254

SECTION VII: IT KEEPS ON GOING...AND GOING ...AND GOING: 1990s

Opportunity/Challenge/Growth 263
A Good Life 364
Ice Cream Makes People Feel Good 264
Secrets of Success 265
Combination Printing = Greater Growth 265
Stop Thief 268
Working for the U.S. Government 268
The Multiplier Effect of Leadership 269
Computerized Pre-Press 270
EPA Compliance 270
Today's Leaders 271
Smart Presses and Smart People 272
Japanese Competition 273
The Constant is Changed 273

The Canadian Connection 274
International News/S.O.P. PSARL Industry 276
ISO 9000 277
A Different Point of View 278
Full Circle? Maybe 278
The Linerless Label 279
Kindergarten Failure 279
George and the Cookie Monster 281
There Are No Barriers 281
Daddy, Don't Go 283
The Future by Farkas 284
Fasson Air Force 284
Only in America 285
It's Only Bigger, Not Different 286
What's Next? A Lot! 287
The Beat Goes On 288
The Ultimate Label 288
You Are Your Own Three Million Dollar Customer 289
Hold The Printing 290
You Want It, We'll Make It 290
On the Curve 291
Tip of the Iceburg 292
The Whole Evolution 293
The Past Is Prologue 296
Chronology of Applications, 1990s 297
Chronology of Invention/Innovation, 1990s 297
The Man Who Wouldn't Quit, Part Seven 297

SECTION ONE

THE FIRST TEN YEARS:

1935-1945

THE LABEL

The term "label" originally meant a band or ribbon of linen or other material attached to the fillets on a bishop's mitre. From identifying a bishop to identifying products is a long road, with many turns.

Around 1700 apothecary labels and labels to identify bales of cloth were printed in Europe. Wine labels followed sometime during the eighteenth century. But printing by hand on wooden presses (using hand-made paper) is long and laborious and it was not until the paper-making machine was invented by Nicolas-Louis Robert in 1798 that paper was no longer in short supply. In that same year Alois Senefelder developed the lithography process and by 1830 labeling was common. Colored labels appeared in the 1850s.[1]

The first labels for product identification in the United States, according to Jim English, were apothecary labels, hand made in 1862.[2] His choice of date seems even stronger when one reads the history of Salem Label Company, the oldest label company in the United States. Labeling in the United States began in 1862 when Josiah Mitchell, a young immigrant drug clerk from England, made gummed labels at night for use in his day job clerking in Alfred Wright's store. Josiah Mitchell's night labeling eventually became his business and the company passed through a number of name changes before becoming Salem Label in 1907.[3]

From 1862 to 1935 the identification of products was accomplished through the use of direct printing on a package, carton or container, or by affixing a label by glue, by water gum label or heat seal label, or by fastening mechanically with string, screws, rivets, etc., directly onto the product.

The invention of rubber-based pressure sensitive adhesive is attributed to Dr. Henry Day about 1845. Pressure sensitive adhesives need only activation by pressure to be applied to a surface. Dr. Day's invention was used in the manufacture of surgical and medical cloth tapes. Later these adhesives were used to produce a variety of tape products, including electrical tape, masking tape, cellophane tape. Many of these inventions were pioneered by the Minnesota Mining and Manufacturing Company, or small companies that became part of 3M. All of these tapes were wound upon themselves. They did not have a protective release paper covering the adhesive. These tapes would later be called self-wound tapes.

Other innovations included printed self-wound paper and film tapes. These printed tapes carried company names, messages, instructions. Sometimes they carried prices printed repetitively. The user would tear or cut off a piece containing one or more segments to produce a label-like marking. But the combination of the tape and the adhesive plus the addition of the release paper enabled one man, Stan Avery, to invent an industry.

ONE MAN INDUSTRY

Actually, it would be more accurate to call it a one man, one woman industry, because the start-up money, $100, came from Stan Avery's fiancee, Dorothy Durfee.[4] It was an inauspicious founding of what would become an international Fortune 500 company, and the beginning of the pressure sensitive adhesive label industry.

Ray Stanton Avery was born in Oklahoma City, Oklahoma (then Indian Territory), January 13, 1907. His father, Oliver Perry Avery, had left the family farm in Connecticut in 1891 and headed west. Stan's father was a Congregational minister and Stan's brother Perry followed his father into that profession.

Avery attended Pomona College in California, majoring in psychology. When his professors insisted in calling psychology a science, Stan rebelled and was finally granted a degree in "Cultural Synthesis."[5]

In 1929 Stan and a small group of his friends decided to skip a year of college and "see the world." They formed the "Oriental Study Expedition of Students from Pomona College," and, after pooling their summer earnings, sailed in steerage aboard the Japanese ship, Korea Maru. They visited China, Inner Mongolia, Korea and Japan, and Stan has credited his experiences on the trip with establishing the philosophy of business ethics he would follow throughout his career.[6]

DEPRESSION

Stan and his friends returned to the United States during the Depression. Los Angeles had more than 100,000 unemployed. Businesses failed; millions of dollars were lost.

Avery was luckier than most. He was able to work as the night clerk of the Midnight Mission in Los Angeles and complete his final year of college. In David L. Clark's The First Fifty Years, the official Avery Dennison International Corporation history, Avery recalled his unorthodox solution to purloined sleeping clothes. He made them out of pink gingham, which not a man "would want to be caught dead in." The theft of the garments ended.[7]

After college he worked at the Los Angeles County Department of Charities until, in 1933, he left to become production manager of Adhere

Paper Company. One of the company's owners was fellow Oriental Study Expedition member, Don Defer. Avery's first assignment was to build a machine to create self-adhesive paper sheets that would have a removable protective covering. The original purpose was for sticking funeral signs on car windshields. As would later prove common in the pressure sensitive industry, Adhere first took orders, then tried to figure out a way to manufacture the product.[8] The production machine became Avery's job.

The first such apparatus was unimposing. A roll of paper 26 inches wide was unwound under an adhesive-filled box with two holes in it. The adhesive dripped onto the paper. To dry the adhesive Avery opened the windows in the room. Then the paper passed over a roller where a removable backing paper was applied, and the sheets were cut to size by hand.[9] These first hand-made sheet or "laid-on" labels would evolve into the pressure sensitive adhesive roll label.

Another product was "Gum Tacks," designed to be used to stick advertising signs on windows and bar mirrors. Gum Tacks were 3/4 by 1/2 inch pieces of paper with adhesive coated on both sides, which were then covered by Holland cloth, a starched bonding fabric originally utilized for cloth book covers.[10]

ONE-SIDED GUM TACKS

The Adhere Paper Company had economic problems and failed, but Avery continued to experiment with Gum Tacks on his own. He knew they worked but were extremely hard to handle. To use them, one removed one of the cloth backings. No problem. To remove the other backing, however, it was necessary to hold the exposed sticky side.

Avery abandoned the double-sided aspect of Gum Tacks and created a viable pressure sensitive product, Kum-Kleen.[11] These 3/4 inch diameter circles of unprinted white paper were coated with adhesive on one side and attached to a backing sheet. They could be written upon, peeled off the sheet and applied to almost any surface as price markers. Because the adhesive was relatively weak, they could be removed at the cash register or by the customer at home. Back then, the labels available were "gummed labels," paper strips with a dry adhesive. They were applied by moistening the back and were cheaper than Kum-Kleen Adhesive Price Labels, but harder to remove. They were also more likely to fall off when the label dried than were Stan Avery's labels.[12]

ANGEL IN THE LAB

For ten dollars a month, Avery rented space in the San Lorenzo Nursery of Los Angeles from another college friend, Joe Shinoda. And not just any space. His laboratory was a loft atop a huge floral refrigerator.

Perhaps to help pay his $10 monthly rent, Avery also worked from 4 AM to 11 AM, packing flowers for the nursery. Then he was free to work on the machine to make his pressure sensitive price labels. To get to the loft he had to climb a stepladder.

Again, from an interview in The First Fifty Years, Avery said, "Joe's employees had a favorite standing joke with me....They tied a big bow out of ribbons to look like angel wings. When I would start walking up the stairs to my loft, someone would quietly hang the bow on the collar of my coat. It looked like I was ascending to heaven."[13]

The machine he built was far from heavenly. With a male and female die, it punched two circles of adhesive-coated paper at a time and transferred them to a backing sheet. A complete sheet was ten-up (contained 10 labels). Each sheet was then fan-folded by hand into cartons of 1000 Kum-Kleen circles each.

Kum-Kleen Products was founded in June of 1935. Stan Avery and Dorothy Durfee were married in August. Dorothy proved to be an astute direct mail marketer. A retail store mailing containing a mimeographed letter and ten Kum-Kleen samples was tested and—even though retailers were going bankrupt in alarming numbers—the response to the offer of 1000 labels for $1.00 was excellent, more than ten percent.[14]

Dorothy then targeted drug, hardware, furniture and other retail stores. Although many were Mom and Pop operations, not all were. Kum-Kleen labels appeared in some very upscale retail establishments and the customer base expanded geographically as well. In six months the firm had gross sales of $1,391. It also had its first non-family employee, Joe Old, who had worked with Avery at Adhere. Joe earned $15 a week, Stan, $20.

Joe and Stan soon parted and Avery hired his best friend at Pomona College, Eliot "Digges" Graves.

NEW PLANT AND FIRE

In April of 1936, Kum-Kleen Products moved from the top of the refrigerator to a storefront location on East Fourth Street. Soon Avery leased the adjacent building. The name was changed to the much more corporate Avery Adhesives, and by 1938 sales were averaging more than $2000 a month.[15] Still, Stan was unhappy. The product was not as good as it should have been. The slow punching operation left adhesive around the edges of the circles (and also the squares and rectangles they now made).

The solution was simple in concept but would prove extremely difficult in execution. And, as orders kept coming in, there was no time to experiment. In late October of 1938 the solution to the time problem came from a most unlikely source—a disastrous fire that destroyed the machinery but left untouched the inventory in the other building.

Digges Graves had been mixing a batch of adhesive and there was a major solvent spill which they covered with sawdust. Hours later when they returned from dinner, the room seemed clear of fumes, so Digges lit the pilot light on a gas heater. It could have been his final act on earth. Hair and eyebrows singed, Digges fled the room, and soon the solvent exploded.

Stan Avery may not have ascended into heaven by stepladder, but someone up there was surely looking after him. At his father's insistence (he had lent his son $1,000 for the business), Stan Avery had taken out fire insurance. The insurance covered all but a few dollars of the loss. Rather than rebuilding with the old technology, Avery decided to make the big jump to precision diecutting, creating today's pressure sensitive adhesive business.

HEART OF FLEXOGRAPHIC

Although there are many exceptions, the pressure sensitive adhesive roll label industry today generally involves something printed on a pressure sensitive adhesive-coated material, usually on a backing or release sheet, and printed, most often by a flexographic process rotary press.

In the early days flexographic printing was a crude process not much advanced beyond printing with a rubber stamp and an ink pad.

The first breakthrough came in 1939 when Doug Tuttle, vice president of marketing for International Printing Ink Division of Interchemical Corporation, invented the anilox roll. The company later became Inmont for a time, then a division of General Printing Ink.

Joe Trungale of Pamarco calls the anilox roll the "heart of the flexo press because it is a means of metering or measuring the amount of ink in the reservoir and transferring it to the printing plate. It regulated that amount of ink through a series of pockets, holes or cells that are engraved across the entire surface of the roll."[16]

Trungale, who would supply such rolls to Mark Andrews, Sr., in the 1940s, describes the process before the anilox roll:

> *[In] the early days the ink distribution system was one that was transferred through two rubber rolls similar to an old wringer washer. And the ink's volume was controlled by the amount of pressure exerted through these two rolls. The process...was called aniline printing mainly because the ink was composed of aniline dyes....The word anilox comes from aniline and the "ox"—as strange as it seems—[came from] a brand Interchemical was marketing...called oxygenated inks. In the litho industry they called it lithox. [The anilox roll was made of steel], mechanically engraved and then chrome plated....It replaced the top rubber roll in the ink distribution system. So the rubber fountain roll picked up the ink, transferred it to the anilox roll and the anilox roll transferred it to the printing plates.*[17]

The early anilox rolls probably had 150-160-180 cells per linear inch, says Trungale, so an area "the size of a postage stamp might have somewhere in the neighborhood of 30,000 to 40,000 cells."[18]

Today's laser-engraved rolls have cell counts of 800 or more per linear inch, but the anilox roll remained relatively unchanged until the 1950s.

INNOVATIONS TO 1940

Avery's first adhesives were made from natural rubber and, like all natural rubber-based adhesives, were inconsistent. Sometimes they stuck entirely too well. Sometimes they didn't stick at all. So Avery traveled to the Los Angeles office of B.F. Goodrich and asked to see a chemist. Jack Lockridge was the chemist and he told Stan about a new synthetic rubber called "Vistanex." Avery commissioned Lockridge to develop a synthetic adhesive for a fee of $25![19]

The chemist returned with two samples, one of Vistanex, the other using natural and synthetic rubber. Vistanex proved superior to the natural compound in standing up to heat and light. Shortly thereafter, Lockridge left Goodrich and joined Avery as chief chemist.

To create a die that could not cut through the backing sheet Avery created a successful but very expensive die from a single piece of steel. Then he hit upon the idea of using a steel watch spring supported on its edge in a metal plate. When the spring became dull, it was easily replaced.

The first release liner was also a problem—it was very inconsistent. Stan set about trying to improve it. Castor oil had sometimes been used as a lubricant in drums of Vistanex and it worked very well. Avery experimented with various ratios of castor oil and lacquer. This led to the combination castor oil/lacquer release coating that served the Avery business well for twenty years.

Other early Avery inventions included a heat process to release the matrix of adhesive-coated paper and a label dispenser developed for Berkshire Knitting Mills. A wooden matchbox was his inspiration for the design.[20]

NEAR MISS

Don Buchta, of Mid-America Tag and Label Company, started his career in the research lab of Marathon Corporation while still attending college on the GI Bill. Marathon was the largest user of paraffin wax in the country in those days. While in the labs he discovered that Marathon had developed a pressure sensitive material, though lacking a backing paper, in approximately 1938. After learning of Avery's patent, the giant chemical company discontinued development.[21]

TANGLE WITH THE BIG BOYS

Avery's dispenser patent application ran into problems because it conflicted with a similar patent held by Dennison Manufacturing Company, at that time the number one name in gum labels. While Dennison did not sell a pressure sensitive label, the company felt large usage of the upstart technology would be in production lines, so it took out the dispenser patent to protect itself.

Stan Avery went to Dennison in Massachusetts where the big fish offered to swallow the smaller fish. Stan said he would agree to be purchased for a quarter of a million dollars, but was told Avery Adhesives wasn't worth it. As he later said, "And they were right."[22]

Finally, an agreement was reached. Dennison won the patent (by three months), but agreed to let Avery use it without a fee. In return Avery would produce "Pres-a-ply" labels for Dennison.

Would Dennison's great size (they had 30 salesmen to Avery's one) consume all of Avery's still-limited production capabilities? It never happened. Dennison's sales staff, fearing the loss of their bread and butter gum label sales, never really pushed Pres-a-ply.

By 1940 Avery Adhesives sales reached $71,857.[23]

THE WAR

When the Japanese captured the rubber plantations of southeast Asia, it was obvious that both natural and synthetic rubbers would soon be in short supply. Avery bought what he could and soon could get no more.

A government official in Washington told Avery his products (such as cosmetic labels and price stickers) were certainly not priority items, and that he would receive no more synthetic rubber. From The First Fifty Years, Stan is quoted as saying: "I was out of business if that decision stood. At that point, I literally broke down. I cried.... That was probably the company's lowest point."[24]

Avery was able to circumvent that particular bureaucrat, and continued in operation, with much of the company's production war-related. Industrial plants used pressure sensitive labels to put safety warnings and operating instructions on machinery. "Kum-Kleen" stickers and labels filled in for metal tags. Avery even manufactured labels with the Morse Code alphabet for radios to be dropped into the ocean for shipwrecked sailors.[25]

The war affected Avery in a significant way. Although they had had some early industrial experience with Standard Oil, Max Factor (silver and gold foil) and Schaeffer fountain pens (transparent film labels), Avery Adhesives was primarily a retail price marking operation before the war. Only 15% of Avery's pre-war sales were in industry. By 1945 85% was industrial.[26]

INTERNATIONAL AVERY

Maurice Bezencenet, a Swiss manufacturer's agent, was visiting Los Angeles in 1939 and saw the Avery Kum-Kleen labels. He liked them and came to Avery's office. Bezencenet acquired the rights to sell Kum-Kleen labels anywhere in the world outside of the United States.

Bezencenet suggested Avery take out a trademark, but Stan didn't think it was worth the $250 required. It's possible the Averys didn't have $250—Dorothy's job as a fourth grade school teacher ended with the birth of daughter Judy in 1938. So Bezencenet paid the $250, promising to reassign the trademark to Avery in return for the $250.

The Swiss agent was successful with the product and it accounted for a "significant source of income" for the young Avery Adhesives Company.[27]

One thing Bezencenet discovered was the rich potential for sales in Latin America, Africa and Asia, because in high-humidity regions pressure sensitive labels stayed put when gum labels would fail. World War II ended the profitable business and, following the war, Avery went to Europe to end the relationship (the company had completely changed direction during the war).

It cost Avery $20,000 in labels to get his trademark back. It was a $19,750 mistake. But Europe was to play an important role in the company's eventual success.

ESTABLISHING THE MYSTIQUE

In the first ten years of the fledgling industry, Avery Adhesives became the prototype for most of the PSA (Pressure Sensitive Adhesive) companies that would be established over the next two decades. The criteria include:

1. The PSA industry is entrepreneurial.

2. It is a family business.

3. The initial investment is low.

4. It will target a fairly narrow niche in the marketplace.

5. The entrepreneur will be innovator, salesman, marketer, manager, and work 60-100 hours per week to establish the business. If one of those skills is lacking, a family member or old friend will provide them.

6. The company will be plagued with problems of quality control, consistency and delivery.

7. The business will grow rapidly and prove exceptionally profitable but <u>not necessarily along the lines the entrepreneur first imagined.</u>

8. Customers will drive the company. Later in the company's development, salesmen—or salesmen collaborating with customers—will create new challenges and opportunities for the company.

9. Companies will be highly secretive about chemistry and production line techniques.

10. Suppliers will supply technical help in return (they hope) for additional business. The seeds had been sown that would allow the incorporation of technology from other, more established industries.

11. At some point, if the company is successful enough, the entrepreneur will seek expert help in the field(s) in which the entrepreneur believes he is lacking.

CHRONOLOGY OF APPLICATIONS, 1935-1945

1935	Blank Price Marking Labels
1938	Promotional Labels
1942-45	Name and Serial Plates
	Inspection/Information Labels
	Cosmetic I.D. Labels
	Secondary Product I.D. Labels

CHRONOLOGY OF INVENTION/INNOVATIONS, 1935-1945

1935	First Pressure Sensitive Adhesive Label on a Liner
1939	Anilox Chromed Steel Roll
1940	Roll Label Dispenser Box
1942	Improved Release Liner
	Synthetic Pressure Sensitive Adhesive

THE MAN WHO WOULDN'T QUIT, PART ONE

Harold Scherer started in business during World War II with a company that made hydraulic controls for blast furnaces, and:

> *...after the war some friend of the president of the company came in and he was just a jerk. And no one liked him and I just decided that life was too short to work with someone who you didn't enjoy working with....The vice president...Mike Voss...wouldn't hear of my leaving....I said no...I won't leave you in the lurch...To make a long story short, I stayed there... on a project I was working on...and they gave me time off to look for a job....There was a very small company...had around nine employees....I went to work there shortly.*[28]

The nine-man company for which he interviewed was Kleen-Stik. Harold Scherer had moved to the PSARL (Pressure Sensitive Adhesive Roll Label) industry.

[1]FINAT, The Self-Adhesive Label: Progress and Innovation. Turin, Italy: G. Canale & C.S.P.A., 1988.

[2]Jim English interview with Bill Klein, p. 9.

[3]Salem Label Company article, c. 1968, p. 1, 2.

[4]Clark, David L. The First Fifty Years: Avery International Corporation Fifty-Year History, 1935-1985, Los Angeles: Avery International Corporation, 1988, p. 16.

[5]*Ibid.*, p. 8.

[6]*Ibid.*, p. 8.

[7]*Ibid.*, p. 9.

[8]"Stan Avery: Father of the Pressure Sensitive Label." Narrow Web Journal, April/May 1990, p. 17.

[9]Clark, The First Fifty Years, p. 14.

[10]Don Buchta interview with Bill Klein, p. 2.

[11]Clark, The First Fifty Years, p. 16.

[12]Today a much refined "gum tack" product is marketed world wide under the trade name of Magic Mounts, produced by Miller Products in New Philadelphia, Ohio.

[13]Clark, The First Fifty Years, p.16.

[14]*Ibid.*

[15]*Ibid.*, p. 21.

[16]Joe Trungale interview with Bill Klein, p. 1.

[17]*Ibid.*, p. 2.

[18]*Ibid.*, p. 5.

[19]Clark, The First Fifty Years, p. 26.

[20]*Ibid.*, p. 28, 29.

[21]Buchta interview, p.2.

[22]Clark, The First Fifty Years, p. 33.

[23]*Ibid.*, p. 34.

[24]*Ibid.*, p. 36.

[25]*Ibid*,. p. 40.

[26]*Ibid.*, p. 41.

[27]*Ibid.*, p. 92.

[28]Harold Scherer interview with Bill Klein, p. 1, 2.

SECTION TWO

COMPETITION STIRS:

1945-1955

THE SECOND FATHER

Even before World War II, one of the employees of the Curtiss Wright corporation in St. Louis, Missouri, had a product identification problem. Aluminum was entering the plant in many varying thicknesses, 016, 020, 024, etc. The thicknesses often got mixed up in assembly, and the finished product or sub-assembly had to be rejected.

The man in charge, Mark Andrews, Sr., hit upon the idea of marking the sheets with a rotary marking device. He obtained a device for marking cartons from S.G. Adams Company in St. Louis. It had a handle, a roller and a rubber plate. When ink was applied to the felt roller it inked the rubber plate, marking the carton. Andrews modified the device so that he could mark the twelve-foot wide aluminum sheets with an image. Using a rapid-drying ink, thickness and other specifications were marked about every foot. Rejections stopped.

In a reminiscence taped two years before his death in 1980, Mark Andrews, Sr., says, "I think this is really the first narrow web flexographic printing unit I can brag on."[1]

Soon after, Andrews created a similar device to mark metal tubing for Curtiss Wright and, near the end of World War II, for Emerson Electric. In Mark Andrews' own words,

> *I had developed some moveable type, which I made by gluing a strip of letter characters onto a rubber base, and then I cut them apart on the lathe, and then made a V-shaped groove...in my printing roll that I could insert those round rubber pieces in. Then I would cut those pieces apart so I could compose my own text.*[2]

Irwin Huber, Sr., a printer in York, Pennsylvania, saw Andrews' machines and bought one for $15.00. Huber had developed a device for wire marking. It involved a series of disks with engraved numbers, a ring of 0s, a ring of 1s, a ring of 2s, etc. Huber placed them on a spindle, inked them and then ran a transparent tape over the spindle so the numbers would appear on the adhesive side of the tape. Then Huber laminated a white opaque tape to it, sealing the printing between two pieces of tape, and wound it into rolls approximately 3 inches in diameter. Unfortunately the rolls would telescope badly in use. Huber gave Andrews an order for moveable type.

Soon Andrews received a frantic call from Huber, who had sold a machine to a Dodge Motors plant in Detroit making aircraft harnesses. "Don't make any more type—it isn't working," Huber said, according to Andrews.[3] The Dodge plant had spent four hours setting the type, but when the adhesive side of the tape touched the type it pulled all of it out of the case.

Andrews looked at the problem and suggested Huber print on the white, rather than the transparent, tape. This called for a new machine which Andrews offered to build. Since this occurred near the end of World War II when metal was unobtainable, Andrews used either hard plastic or hardwood. In his own words:

I bought an endless belt of fabric that I guess was about ten inches long...and I ran my tape over a slotted roll, little slots you know, around the roll, on which I later got a patent, only to discover that 3M had done it that way years before and had decided that the knurled pattern on the roll was better than the slotted pattern. Well, I ran my tape over that slotted roll and then down to the rewind, and the fabric belt slipped in my V-groove on the rewind so that it acted kind of like a clutch. I had a little pad that I could move to increase or decrease the tension on the fabric a little bit, and lo and behold, I had a clutch rewind which rewound the tape and it didn't telescope. Well, I took the machine up to Mr. Huber and he liked it so well he said, "Make me ten of them. How much would they be?"[4]

Andrews was commissioned to provide ten machines at $300 each. Only one was delivered, but it did meet all but one of the criteria (liquid ink) for pressure sensitive flexographic printing. Soon, on his own, Andrews created a true flexographic unit to print on water moistenable gummed tape.

It worked, but there was little demand. Still, in 1946, Andrews left Emerson and went into business on his own. His first large order came from Swift, which was having a promotion on packaged meats.

We ran it out of two inch wide...tape with a two inch wide transparent cover over it. And I think the order was for something like 2,000 rolls. These were small rolls...but it was a lot of tape and a lot of money. So I called the 3M people and asked them if they could give me some idea of what sort of credit I could have on an order like this. The salesman said, "Are you likely to spoil any tape?" I said, "If I spoil any, I'll really go bankrupt." He said, "If you don't spoil any...why, we'll finance the whole deal for you. We'll give you credit until you can pay us from getting paid from Swift."[5]

The profits from this sale, plus $6,000 from his father, staked Andrews to working capital. His uncle Frank came to work with him. Finally, about 1947 as he recalled, 3M developed a printable self-wound tape and then, "we were off to the races."[6]

By the end of World War II the 3M Company recognized that the printed tape market was an excellent outlet for their tape products. The company put together a marketing program and hired a field sales organization to promote their tapes for printing. This proved to be a highly successful program. The use of printed self-wound pressure sensitive tape spread rapidly throughout the United States and into many industries from food packaging to electrical appliance assembly.

As a sidelight to the history of this industry, Andrews remembered a "fellow by the name of Vickory in Peru, Indiana, [who] apparently could print on any pressure sensitive tape." Vickory's advantage, before the 3M development, was probably a breakthrough in ink formulation, but the man died and his secret was lost.[7]

The printable tape was of acetate fabric with some body to it, leading to the idea of perforating the printed tape. "So," said Andrews, "we fixed up a perforating device and, lo and behold, we could perforate the Scotch Tape and we sold a lot of [it] because it had a nice workup and the printing was centered properly between the cuts so it had the reasonable appearance of a label."[8]

The original basic tape printers produced by Mark Andrews were sold for under $1,000. Thus a person could go into the tape printing business for perhaps as little as $5,000, including first month's rent on a small shop, a supply of unprinted tape, and "eating money" to get the business off and running. In 1950, Andrews would sell an improved tape printer, with the perforating and slitting mechanism and eleven plate rolls in three inch to 18 inch repeat sizes for $1,800. A 25 percent deposit was all that was required, with the balance to be paid out of tape printing profits over NO specified period of time.

The early Mark Andy presses were capable of printing only very narrow widths of tape—two inches, three inches or four inches—as was the first press designed to print and die cut pressure sensitive roll labels. This began the "narrow web" flexographic segment of the label printing business. The striking advantage of this press was its simplicity and very low cost. The disadvantage, on the other hand, was that it was capable of producing only very simple one or two color labels. This was a greatly limiting factor, since far more sophisticated applications were being produced by Avery Label and by a number of the old line tag and seal companies who had offset and letterpress printing equipment.

CHRISTMAS TAPE

"I started work in the shop as the person rolling Christmas tape. They had a special order where they rolled five different pieces of tape on a single core and they sold this for decorating."[9] At seventeen, John Garber got his first job with Mark Andy as the temporary tape roller.

> *There were two companies in that building at that time. One was called Mark Andy. One was called Mark Andrews Company. The Mark Andy Company was strictly involved in the printing and converting part of it. The Mark Andrews Company which was on the other part of the wall...shared the same facilities...same lunch room and break room....The Mark Andrews Company made the converting machinery....That's where I started in the business....they offered me a job in the Mark Andrews Company working as an apprentice machinist/assembler person....I actually put the machines together that were headed out the door.*[10]

Garber went on to work for a firm called A True Gentleman Company, in Sandusky, Ohio, producers of shipping tickets for the auto industry. He later returned to work for Mark Andrews, then went on his own as a specialist, setting up converting companies and getting their line running.

A BUSINESSMAN JOINS THE DREAMER

When World War II ended in 1945, Avery Adhesives was poised for explosive growth. The war had changed the Stan and Dorothy Avery partnership. There were now 50 employees, but a number of ex-employees had already set out on their own as competitors in the pressure sensitive label business. Among the new competitors was his good friend from Pomona, Digges Graves. Graves, the Avery Adhesives sales manager, and the company's only salesman had formed Contact Products. Stan Avery was crushed. In the mid-1950s Contact sold out to Monarch Marking Systems.

Yet Stan returned to the ranks of Pomona College for his next key management hire, Russ Smith. Smith had graduated about four years later than Avery, but his older brother, Elden, was a friend of Stan's.

Smith graduated in 1936 and worked briefly on Wall Street as a security analyst. Laid off in 1937 (which he later said was a blessing), he went off to Europe to bicycle the continent. He ran out of money by the time he reached Geneva, Switzerland, but was told that if he could learn French quickly enough he could work in the economic section of the International Labor Office, a part of the League of Nations' International Labor Organization. Smith learned quickly and began work just as his total assets dropped to five dollars.[11]

He worked in Europe for three years until the fall of France in May, 1940. Then Russ returned to America where he worked in labor relations with the Blue Diamond Corporation, the largest building materials firm in the Los Angeles area. From 1940 to 1943, before serving in the Navy, Smith represented the construction industry in labor negotiations.

After the war Smith resumed his friendship with Avery and suggested that it was now time to incorporate Avery Adhesives. Stan agreed, realizing how easily his hard-earned business could be lost in the difficult days of rapid expansion. Avery says he called Smith and asked him, "How'd you like to have my job?"[12]

Russ Smith became secretary-treasurer of the newly incorporated firm, and on March 15, 1947, vice-president and general manager.

William L. Gilman, then company comptroller, is quoted in The First Fifty Years as saying "Russ was the businessman and Stan was the dreamer. They needed each other. Stan had the imagination, the dream, and Russ made it reality and kept it businesslike."[13]

One of the first things Smith did was to write a position report describing Avery Adhesives at that time. The war had changed the company completely in less than five years.

THE THREE MARKETS

According to Smith, business came in three categories:[14]

1. **Resale**. Labels in this category were generally unprinted and sold in flat packages or dispenser boxes. They were sold through 1200 retail stores (mostly stationery stores) and through 300 wholesalers and distributors. Still important, this market now (1947) accounted for only 10% of the sales.
2. **Industrial**. These were sold to manufacturers and applied to products, usually on an assembly line. At this time, most were labels for consumer items and could be plain, printed or embossed. Such sales were 80% of total Avery sales in 1947.
3. **Converter**. This came from sales of pressure-sensitive materials to printers and manufacturers, such as distributors of price-marking machines sold to large retail stores and chains. Converter sales accounted for only 10% of the total in 1947.

Avery had sales offices in Los Angeles, Chicago and Detroit, and distributors in New York, Boston and Pittsburgh. Two dozen sales representatives sold Avery products in most of the largest U.S. cities. In 1947 the average invoice was only seventy dollars but there were about 3,000 active accounts in almost every industry and trade.[15] Founded in the Depression, Avery was now almost recession-proof because if one type of business stagnated, other markets grew to replace it. Russ Smith felt this was a major strength of the organization.

A SWISS MISS

Although Avery by now had almost fifteen years of success, the Swiss were wary. In the fall of 1950 Ferdinand Rüesch of Gallus met Stan Avery at a graphic arts exhibition in Chicago. Avery liked the idea of a Swiss connection and offered Rüesch an Avery agency for Switzerland. At the time this meant pressure sensitive material, label presses and label construction. When he returned to Switzerland Rüesch spoke to his father, the president of the company, about Avery's offer. His father replied that "they should be rather careful, and that in connection with Chicago you could hear so much about gangsters that the Rüesch company should rather stick to their honest business and watch how this Avery company would develop before taking such important decisions."[16]

3M

Others tried to compete in the pressure sensitive label business in spite of Avery's patents in the field, but it was both difficult and expensive. It was also a lot of work for a $70 sale.

These firms did not produce the pressure sensitive adhesive coated base material as did Avery Label, but purchased these materials from 3M, Kleen-Stik or Simon Adhesives, the only firms producing pressure sensitive material with a release liner.[17]

Minnesota Mining and Manufacturing, the 3M Company, was already a large business, and apparently felt that pressure sensitive labels were small potatoes. Scotch Tape was their baby, and if someone wanted to print on it or die cut it, fine. The irony is that many future pressure sensitive adhesive roll label companies were founded by 3M salesmen who could sense the great potential even if their company did not.

One reason the 3M company failed to capitalize on this leadership in tape and adhesive technology was their unwillingness to customize their product to meet the needs of (mostly) printers who wanted to produce pressure sensitive labels. John Orlando, who was with Simon Adhesives for 16 years, recalled 3M would offer a customer product but would not substitute an adhesive or backing sheet to accommodate what was to 3M a ridiculously small order.[18] Of course 3M has prospered, but it never did become a major player in the pressure sensitive roll label business.

PARK AVENUE PLAYBOY

Simon Adhesives Products made Easy Stick. Simon Adhesives was founded by Sander and Mark Simon, but in 1950 the brothers had a falling out and Mark left the company and founded Pressure Stick.

Sander Simon was a unique individual. John Orlando, who worked for a number of pressure sensitive pioneers, recalled his former boss with much affection.

> *He taught me everything I know....He taught me integrity, honesty...that have governed me over the years. He was a wonderful teacher...a fabulous, fabulous person...a man of his word.*[19]

He was also a playboy:

If you talk to other people in the industry they'll tell you about Sander being a playboy. As a businessman, he wasn't....He lived high on the hog....He used to live on Park Avenue. He dressed the best, he lived the best, he ate the best.[20]

Dick Capuzzo of General Trademark recalled visiting the Simon plant in Long Island City as a child. His father took him there on a Saturday:

I can remember the noise. The place was noisy as anything. [And it] smelled like hell. But we met the owner who was Sander Simon. He was a real character. Drove around in a Rolls-Royce convertible. He filled the role [of playboy] so well. He was a nice guy, though.[21]

Capuzzo recalled a problem Simon had:

There was a point where they almost went under. They started manufacturing materials and somehow didn't put a barrier coat on the paper before applying adhesive. What happened was the adhesive soaked right into the paper and dried out. You would wind up with two dried pieces of paper, a liner and a top sheet....It had to be thrown out. But somehow enough people [customers] hung in and he was able to pull himself out of the hole and actually produced real good products after that.[22]

SHOOTING LABELS

Both Monarch Marking Systems in Ohio and Soabar in Pennsylvania had developed a semi-automatic labeling gun. It used Simon Adhesives Products' earliest roll product. John Orlando recalled Simon's original coater was 13 inches wide and it produced a "patterned adhesive....We were the first to have two levels of adhesive on the same web...and we enjoyed all of Monarch's business."[23]

For his first release liners Simon used a nitro cellulose coating. "We worked closely with Riegel...[who was] producing glassine...and they helped us."[24] Simon purchased adhesive from a company named Paisley.

Simon Adhesives Products (and Kleen-Stik in Chicago) started out with a product similar to the one that Adhere had made earlier, pressure sensitive strips to affix advertising signs, and later sheets for printers to print advertising point-of-purchase materials. This was called strips and spots. Sander had some very large accounts (including Coca-Cola) and produced one product that almost everyone can remember—the oil change sticker tags, called Door Jam stickers, which served as a reminder to change oil when the car reached a certain mileage. Robinson Tag did almost all of the printing for Simon, which is where John Orlando worked before joining the adhesive firm. In 1954 Orlando became a vice president, and stayed with the firm until the operation was acquired by Fitchburg, part of Litton Industries, and was moved to Moosic, Pennsylvania.[25]

"HELLO, MY NAME IS ____________________"

Art Schwartz, who joined Kleen-Stik in 1950 to run a new plant in Newark, New Jersey, remembers that the original product had been made by a San Francisco druggist who applied rubber cement to a surface and covered the adhesive with Holland cloth, earlier used as the backing for Adhere's Gum Tacks. Schwartz says they ran the product around the room to dry it.[26]

Gerald (Gerry) Cole learned of the process and before World War II he decided this would be the basis of his business. He started Kleen-Stik, but military service interrupted his career almost at once. Somehow his father was able to keep the business going until the younger Cole returned to Chicago. About a year later, some time in 1946, an Army friend, Wayne Lasinski, joined him:

> *At the time we were just producing sheets, manufacturing transfer tapes for application to printed advertising pieces. They were manufacturing the tape and also applying the tape mechanically in their own plant. At that time they were making sheets, pressure sensitive stock, primarily one type of paper and [coated on] one side.*[27]

Lasinski says they were making stock for roll label converters by 1948. Technically, since neither Kleen-Stik nor 3M printed or die cut labels they were not in direct violation of Stan Avery's basic laid-on-label patent. But they were selling to those who did, including Al Steigerwald in Chicago and the Ever Ready Label of New York. Ever Ready would later become the prime target of the best lawsuit a man ever lost.

Kleen-Stik's best-known product (and probably still the best-known PSA product today) was the "Hello, my name is__________" label. It worked fine on most clothing materials, but it ruined many an expensive suede jacket. When penny products ruin $100 jackets, there are many unhappy people.

CALIFORNIA DREAMING

The biggest development of 1948 came, not surprisingly, from Avery Adhesives. Russ Smith knew it was time to make the big move. In March of that year Avery consolidated its scattered manufacturing and shipping operations into what was then considered a large plant (17,000 square foot) at 1616 South California Street in Monrovia, California. It was the first company-built, company-owned factory in the company's history.

Stan had spent a lot of time designing the new factory's showpiece, a 26 inch coater. But it just wouldn't work. In an interview, Russ Smith said:

> *The 26-inch coating machine was a monster. The engineers at the Monrovia plant wrestled with it and wrestled with it. There was always some reason why it didn't produce.*
>
> *To meet the orders coming in, we kept the 8 inch machine going at the old plant on Union Street in Pasadena. Finally Stan and I said, "There's only one way that new machine is going to run and that's if we tear down the Pasadena plant." So that's what we did. The new machine ran beautifully thereafter.*[28]

In the beginning the goal had been to create a product that could be applied easily and then removed easily as well. When Korean War rubber shortages forced Avery chemists to develop new adhesives, they developed what turned out to be a permanent adhesive, far different from the removable or Kum-Kleen adhesives in use. Avery called the product Perma-grip and it was colored blue as a warning that these labels, when applied, were there to stay. Almost instantaneously, half of the business utilized the permanent adhesive. In retrospect, it is surprising this adhesive was not developed earlier, especially during World War II.

Avery held nineteen patents including the original one for producing self-adhesive labels, "Method and Apparatus for Making Label Units." According to The First Fifty Years, this was a patent on a method, not on a product, because self-adhesive products such as Band-Aid® bandages and corn plasters preceded 1935.[29]

In 1952 Avery brought suit against Ever Ready Label Corporation of New York for patent infringement.

THE WHOLE BALL OF WAX

In 1952 Ever Ready Label was perhaps the largest label supplier (primarily gum labels) in the United States. It was certainly the best known in New York City. The story goes that Ever Ready's Sidney Hollaender went to the New York telephone company and suggested they create a book that listed firms by industry and publish them in a separate phone book, in a different color.[30] For his invention of the Yellow Pages, says Harry Stillman, who joined the firm in 1951, Ever Ready received a 99-year lease for the rear cover of the Manhattan Yellow Pages at one dollar a year. One ad featured 5000 gummed labels for 99 cents.[31]

The company's founder was Sidney Hollaender, Sr., a "good old boy. He would never ride an airplane. He always rode the train. He'd go to Florida, he'd get on a train. Wherever he went. Like the football announcer, [John] Madden."[32], says Herb Smith.

A dedicated Lincoln buff, Hollaender decorated the company's New York office with Lincoln memorabilia, including a "set of gates that had been at Lincoln's home."[33] Hollaender also paid for a large billboard on 42nd Street, featuring the Emancipation Proclamation.

In the 1920s Ever Ready had sent out catalogs of the company's broad line, including what were claimed to be the first round gummed labels in rolls, offered as early as 1914.[34]

When Ever Ready was sued by Avery, Sidney Hollaender enlisted financial aid from some other companies to help pay the legal costs involved. One of the companies involved was a spin-off, located in the Bronx, called the Alan Hollaender Label Company. Alan, Sidney's cousin,[35] had been sales manager of Ever Ready in the 1940s. After a disagreement, Alan walked out and formed his own company. In a few years, perhaps due to Sidney's untimely death in 1954, the Alan Hollaender Label Company sales soon equaled or exceeded those of the Ever Ready brand. But Alan did chip in with the legal expense money in 1952, as did eleven other companies. Among others, Sidney called Gerry Cole of Kleen-Stik, Sander Simon of Easy Stick and the J.L. May Company of Manhattan; Steigerwald in Chicago; Linear, York Label, Cameo Die and Mold, Top Flight and several others in Pennsylvania. The "war chest" totaled $12,000.

George Collons of Kleen-Stik, who attended the trial in the District Court of New Jersey, recalled that Gerry Cole had tried to convince Stan Avery that "the future of the business was going to depend upon wider application than he [Avery] could produce."[36] Collons said that Cole offered

Avery a royalty arrangement on materials put into rolls, but that Stan "wanted the whole ball of wax."[37]

The court ruled that although Stan Avery had contributed "a step forward," his work involved "the skillful use of human methods and apparatus" rather than a completely new process. It was "an improvement which does not amount to an invention."[38]

NOT MUCH AGONY IN DEFEAT

Avery's Russ Smith recalled, "Stan and I spent Thanksgiving locked up in a hotel room in New Jersey fighting that case. When we lost the suit, we thought that the world was coming to an end."[39]

Stan put it more simply, "It was as scary as hell."[40]

Sometime in the early 1960s, at a packaging show in Chicago, a number of the trial participants were present, including Gerry Cole and Sander Simon. Stan Avery came over and announced that "the greatest thing that ever happened to me was losing that lawsuit."[41]

Why?

Firstly, industry abhors the single supplier. If, for example, General Motors could buy pressure sensitive roll labels only from Avery Adhesive Label Corporation, they would search for another way to solve the problem rather than go with a single source. A strike, a fire, a flood could knock out the one source and cripple GM's production. With a second source, such fears are calmed. Today's computer manufacturers spend millions on a chip, then offer it to a competitor to produce, just to allay such fears. Said Avery in 1993, "At the time, the fact that we had no patent made our customers feel good. They never felt they were locked into a single supplier and they always felt that they were buying at a competitive price."[42]

Secondly, instead of having one company beating on doors hawking the advantage of pressure sensitive labels, now there were many more. Also, in spite of direct sales offices, distributors and representatives, Los Angeles is well removed from America's industrial heartland and the traditional printing centers of the east and midwest.

COATED PRODUCTS

Jim Kyte joined Coated Products in 1954. The company had begun in 1936, founded by Louie Potter and Frank Bennett, two men from Johnson & Johnson. They were making cloth tape which was used as masking tape during the war years. They developed a "masking product which was kind of a label stock—two layer—they had a crepe paper release on it."[43] This was used during the war to put "US Army" and serial numbers on military vehicles.

Coated Products was supplying some label stock in the early 1950s to Soabar and Kimball Systems.

> *They put strips of adhesive down with a finger edge, they could get it off. The labels were all perfed. Soabar had been buying and always bought a lot of stuff from Avery. Soabar was Avery's first big customer in 1936. My father was president of Soabar.*[44]

Some four or five years after the elder Kyte retired, the Soabar company was sold to Avery, "Which is funny," Jim Kyte says, "because in the 1950s Soabar was considering buying Avery."[45]

Kyte remembers something of the famous 1952 patent suit. [Sidney Hollaender] "and Julius May of J.L. May Company, and I think Al Steigerwald was part of it."[46]

Coated Products was one of the early users of the silicone release liner:

> *There were a couple of guys who had products. We had a silicone product and Riegel Paper had a silicone product. Avery, at that time, had a good release liner on a parchment paper, but not silicone. Our stuff we could run a lot faster....That was 1955.... When we first ran silicone, we had to cure it as close to 360 degrees as we could get....We had a hose blowing water back into it.*[47]

Kyte remembers Irwin Huber of Top Flight who made his own rotary dies. Top Flight purchased some of the masking product Coated Products made, so Coated Products worked closely with Top Flight in product and die development. About 1956 Coated Products' owner, Louie Potter, sold out, and Kyte bought the business. In 1960 Huber started another company, Adhesives Research, a current producer of pressure sensitive base materials.

Unlike many of the people in this book, Kyte never worked for his father. "He was vehemently against nepotism. I couldn't even have a summer job there." But Kyte says Potter "hired me...[thinking] that would help him get some business from Soabar. And it did."[48]

Coated Products' first large customer was C.O. Dix, and Kyte remembers some business with Ever Ready, J.L. May and Kimball Systems. Soon Alan Hollaender began to buy in large quantities. He remembers that Avery and Kleen-Stik had "tied up" Patterson Parchment for their total output of parchment for release liners, and they would not sell to Coated Products, so Kyte went to Crocker Burbank and got them to "take some of their semi-bleached kraft and run it through the glassine calender and that worked pretty good."[49]

In the fifties you couldn't get anybody to concentrate on the PSA market—it wasn't big enough. Then all of a sudden this data processing thing started—data processing labels. There were a couple of big guns we were doing business with—Normandy Press and Alan Hollaender...and then Standard Register and that whole group got in and Dow was selling a lot of silicone and getting ten dollars a pound for it.[50]

COMPETITORS, ENTREPRENEURS, INNOVATORS

With the market now free of patent restraint, the number of firms producing pressure sensitive roll labels increased from perhaps 15 or 20 in 1950 to 40 or 50 by 1955. The bulk of these firms had already been in the tag and label business (water-gum and heat-seal types) and found the pressure sensitive roll label business an extension of their capabilities. Now they could produce and/or sell PSA labels to their regular customers, some of whom had become interested in these more costly labels because of their ease of application.

There was one more reason the industry was about to enjoy spectacular growth. In 1949 or 1950 Gerry Cole and Jerry Zalkind of Kleen-Stik had contacted Mark Andrews, the man who could print on Scotch Tape. They asked him to try out some of their pressure sensitive materials. The materials ran well, prompting Andrews to improve his machinery. He substituted a cross score blade for the perforating blade and "put some bearers on the roll that had the cutting blade in it, so the cutting blade would cut uniformly all the time....I would say in 1949 or 1950 we did the first rotary cutting of label stock."[51]

Ironically, the Avery Company had created a rotary die machine during World War II for a large U.S. Navy order. It was designed and built in only two months, but the order had been completed (with the help of a second shift) before the machine could be placed into service.

"NO SMALL POTATOES"[52]

Bob Irwin had been in the industry a number of years before going into business on his own. He had worked for ten years in sales for Larry Black of Transparent Products, and left to join Midway which he sensed was "just about ready to take off."[53]

But when that didn't work out, he started Linear Products in 1952, with a Mark Andy. He knew from the first he wanted no part of the "Mom and Pop business."[54] So he went after the packaging and primary labels for Colgate and other large volume companies, gave them what they asked for.

"I had good people....I wouldn't trade my experience. I would go right back into it again."[55]

"FLOOD" PATENTS

Walt Webster was with Dennison in the early 1950s. He recalled the brand name "Pres-a-ply" for the Avery-supplied Dennison product. He says Dennison's strength was due to the prolific inventions of engineer Carl Flood. Webster says the new line of pressure sensitive labels Dennison offered took off slowly because the sales force was ill-prepared to sell the product and changing a "tag oriented" sales force to push labels was more difficult than Dennison's marketing division had envisioned.

Webster's first sales consisted of good-sized orders "for pressure ply labels which were printed 'number, style (or item) or price'. These labels would be filled in by the hardware retailer. For a number of years...the wholesale hardware dealers supplied small hang tags to their customers...as giveaways."[56] The pressure sensitive labels replaced the hang tags which were cheaper but demanded a lot of the retailer's time and effort. One of Webster's bigger sales in those early years was to Allied Chemical in Hopewell, Virginia:

> *They used a striped adhesive as paper labels that were run through a Dennison Dial Set Printer, designed to handle variable copies. The labels were then applied to the inside of metal pirns [cones or reels for yarn].*[57]

At the time, Webster says, pressure sensitive labels could not challenge water gum, heat seal or stapled tag in soft goods such as hosiery mills "because of cost. Had we been able to supply automatic labeling equipment the volume would have been appreciable."[58]

"I HAD TO SEE HOW THE WHEELS TURNED"

Harold Sidrane of K. Sidrane Incorporated was another of the ex-GI's who got started in the industry right after World War II. He remembers some of the very early players in the pressure sensitive field.

A distant relative had seen a number of people in a company in Indiana who were printing on tape, and suggested this might be a good way to earn a living. The relative mentioned that the ink came from International Printing Ink, so Sidrane contacted the local salesman, Ed Soldese. Soldese gave him a lot of basic information as did IPI's Doug Tuttle, who had invented the anilox roll a few years earlier. Sidrane also talked to the sales manager of Permacel, Miles Oppenheim, who helped him learn about self-wound tape. Oppenheim wanted printed tape so he could sell more tape. The equipment to handle and print the tape was made by Sidrane.

Sidrane remembers that some of the early competitors in New York included Larry Black and Ed Foodie.

Then somebody said, "Why don't you make die cut labels?" I said, "How do you die cut?" So I got a hold of an engraver in the city...Merket and Sons. They engraved the roll for me. [It didn't work] I figured out that if you put bearers on the roller you could make a delicate adjustment and not cut through the liner. Then I had another roll made by Merket and later I found out that Mark Andrews had done the same thing as I did, independently.[59]

Through Oppenheim, Sidrane met Al Sobel who was working for a paper broker.

He started selling stuff for me....He was one of our biggest helpers in getting into the field. He was doing some tape, making it for mufflers, tail pipes. He introduced me to that business.[60]

Sidrane made other machines, perhaps as many as seven. "One of them is still in use!"[61]

The Al Sobel mentioned by Sidrane was another ex-GI who was looking for a job after World War II. He went to work for Leed's Sales Company and was trained to work with self-wound pressure sensitive tape. He went to a plant run by Bauer and Black, who were producing pressure sensitive cloth tape, learned the product and soon was training the sales force to develop pressure sensitive applications.[62]

Then Leed's Sales decided to give up their paper distribution business. So Sobel and three others decided to open their own company. All four men lived in New Jersey and commuted to New York. The four were riding the train on March 15, 1951. They wanted to decide what to call their new company. "I saw this magazine, and I said, 'Why don't we call it Coronet Paper?' Everyone agreed it sounded good. So we filed the papers and we went into business April 1, 1951."[63]

Coronet Paper, in spite of being founded on April Fool's Day, was off to a good start. Sobel said something that almost everyone in the PSARL industry says:

The potential customer provides the impetus. Somebody always wanted something different. And everybody was saying, "We don't do that." And I said, "Well, we'll try." So we tried and we searched out vendors who would do these specialty jobs for us.[64]

An early application was in the electronics industry. A machine had been developed to place resistors onto a roll of tape.

A long line of resistors. It looked like a ladder. But then they needed some mechanism to protect them in shipment. I got a hold of some corrugated and made a three inch wide roll of corrugated with no corrugations in the middle...and built the machine to do that. I sold that product to them...and I sold them tape.[65]

ON HIS OWN

Bill Eiseman, of U.S. Tape and Label Corporation, was one of the early pressure sensitive entrepreneurs. A Captain in the U.S. Army Air Force, Bill was discharged and began working for a large shoe company. His father, after leaving the War Production Board, was also looking for a job and found one as salesman of the printed tape labels Mark Andrews was making in St. Louis. Everything worked well for three months, then the father became ill, later dying from a heart attack. Bill Eiseman took over and did so well he was soon earning more than Andrews himself.

For some strange reason this annoys company presidents. Gerry Cole of Kleen-Stik was quite angry when George Collons out-earned him. John H. Patterson, whom many believe invented modern salesmanship at the National Cash Register Company, once addressed that same issue by pointing out that some portion of every sale was coming to Patterson, and he didn't care how much the salesman made. He added that if a salesman did earn more than he, Patterson would be delighted to pay him in gold in front of everyone in the company.

But Andrews and Eiseman fell out, and Bill Eiseman and a man in the cellophane bag business, Alan Ross, went into the tape label business with a machine built by Pamarco in Roselle, New Jersey. Pamarco had produced a few printing machines for labels that were much bigger than the Mark Andy units.

The U.S. Tape and Label Corporation was started with $10,000, $5,000 from each principal. Ross did most of the selling and Eiseman, the production. Their first storefront office rented for $75.00 a month.

Our first customers were shoe companies, you know, to put a label in the shoes. Shelf strip talkers they put along the shelf underneath where the prices were...different kinds of food products... advertising agencies. There was product identification...and things like that....They came in a roll and a little dispenser. You pulled off what you wanted. They weren't perforated, you pulled down on the perforating blade and made your own.[66]

WAS IT PROFITABLE?

Pretty small stuff, right?

Wrong. In 1950, their first year, Eiseman and his partner did $42,462 worth of business. (Remember, that was in 1950 dollars.) Not bad. The second year grossed $88,144. In 1952 they sold about $124,000 and in 1953 the firm grossed $165,000.

I would call this business almost recession proof because if you get out and hustle, you can really sell this stuff. You're not limited to one [kind of business]. If the shoe business is lousy you can go sell to someone else. It's unlimited and recession proof.[67]

THEY HAD THEIR CHANCE

In the summer of 1952, Bob Klas had the job that a just-graduated college man would die for. He was making a solid $375 a month at Hamms Brewery.

But the job ended in the fall and Klas found a job with Norwestern Standport in downtown St. Paul, Minnesota. They had been working with 3M engineers making flexible printing plates to go around a metal cylinder for rotary printing. Norwestern had a separate business named Tapemark and it was using the "second press ever manufactured by Mark Andy."[68]

After a year and a half, the owner offered to sell Tapemark on such good terms that Klas decided to go into business for himself. Under Klas, Tapemark "struggled through the fifties with our limited capabilities."[69] The company basically printed tapes, especially in what Klas calls "constructional" uses, such as the "old Chung King combination where they put the noodles in one can on the top and...they combined the two cans by putting a piece of tape around the beads."[70]

In 1958 the 3M Company invited about 30 people, including Klas, to attend a Converter Conference at the firm's lodge in northern Minnesota. (Herb Smith, also interviewed for this book, was another attendee.)

That was the first time we got a look at each other. We had a lot of fun, they were a bunch of pretty good guys there. If it did nothing more, it tended to raise the sense of ethics in our industry, because you know we're going to be very circumspective in what we do and we're competing with honorable people out there.[71]

Klas, as did most of the other Scotch Tape printers, moved into printing pressure sensitive labels and wondered why 3M failed to make products for such converters. When Klas played golf with Bob Miller, whom he calls "the most powerful guy in the 3M Company,"[72] he told Miller that 3M was making a mistake:

I said, "You're letting the converting industry slip away from you because 3M is not playing any kind of role. You're not making enough of the materials and enough variety...that are currently being consumed by the old tape printers." And he said, "Well, I hear you, Bob, but here's the way it is. I go in there before the budget committee and want a million dollars for a coater. I can produce about 28 percent return on investment. And the rest of the guys...are guaranteeing returns of at least 35 percent."[73]

So the giant 3M opted out of the PSA roll label base materials business and the pygmies such as Kleen-Stik, Simon and Fasson survived, at least through this "take-off" era.

HELP WANTED

When Mark Donner left the army after World War II, he needed a job. So did a lot of GI's and there weren't that many jobs. A friend of his did get a job at 3M selling Scotch Tape, and he had an idea for Mark, saying, "I've heard of a company in St. Louis that manufactures a machine that prints on tape. Why not buy one of those machines and then I will sell you tape and we both will have a job."[74]

Donner, lacking anything better to do, wrote a letter to Mark Andrews asking for information. He wasn't really thinking of buying a machine, but felt learning about it might be worth some time. There was no response until six weeks later when a truck pulled up with a box from

St. Louis. Inside was a tape printing press for tape two inches wide and a note taped to the machine, appropriately enough, with Scotch Tape. It read, "Turn it on, you'll see how it operates, pay me when you can." There was also an invoice for $600.

So Donner turned the machine on, figured out how to run it and sent Mark Andrews $50 a month for a year, establishing the Rich Mark Company of Seattle in the process. Dale Bunnell of Mark Andy says Mark Donner delighted in telling this story.[75]

Mark Andrews, Jr., recalled the early days of the Scotch Tape printing business. "I can remember rolls and rolls of Scotch Tape stacked up in our garage out back."[76] In 1949 Andrews, Sr., moved from his basement to a small building in Kirkwood, Missouri. There he built a number of tape printers for his own use. Soon "people were coming to him saying, 'Hey, look, this is bigger than you can handle. Why don't you sell me a machine?' I think it was some salesmen from Minnesota Mining who actually decided to get into the tape converting business because they saw the opportunity there."[77]

These were high energy men, often dissatisfied with their current situations. They usually had little in the way of capital resources, but recognized that here was a business opportunity that required only a modest amount of capital to get up and running. They often mortgaged their homes and borrowed a few thousand dollars from family and friends to get started.

Basically they had only one product to sell—printed Scotch Tape. Andrews, Sr., saw both the advantages and the disadvantages of selling his machines to them. Obviously he could make money selling his tape printers, but he was also creating competition for his own printed Scotch Tape labels.

PROTOTYPE OR TINKER TOY

Many of the entrepreneurs who seized the opportunity first to print on a type of Scotch Tape were 3M salesmen. The company was delighted to discover the Mark Andy tape printer and encouraged their salesmen to promote the business. Not surprisingly, they tried to promote it to offset printers.

The offset printers wanted nothing to do with it. The Mark Andy machine was a tinker toy in their eyes and demanded tender loving care to print a salable roll. Paul Styers of Emblem Tape and Label remembers his father getting a two-inch tape machine from Mark Andrews in, he thinks, 1953. For a time Paul's father, Del Styers, moonlighted as a pressman, but

3M was unhappy that its salesmen were competing with their customers, a classic conflict of interest. So Styers quit 3M and founded Emblem Tape, perhaps in September, 1954.[78] Paul Styers joined the firm in 1964 and recalled die suppliers such as Allied Gear and Preston Engravers. He also remembered how difficult it was to use the early narrow web machines:

> *You had to watch these presses like a hawk. [They] were equipped with old clutches that ran on the friction of a belt or leather strap with a spring. And you could tighten that thing up and get more tension on your rewind, or let it go and, in each case, you would change your registration....It was like trying to play a ukulele that was out of tune.*[79]

Styers says printing was "similar to smashing a rubber stamp onto a piece of paper. That was about the sophistication of it."[80]

Ink manufacturers were less than impressed with these upstart companies:

> *You'd call up IPI and say, "I want to buy some ink here." And the guy would say, "Great, great. I'll take your order. How much do you want?" And I'd say, "A quart," and [they] were used to selling 55 gallon drums. So when you told the guy you wanted a quart of ink he was about as excited to do business with you as a case of the measles.*[81]

PRESSURE SENSITIVE ADHESIVE MATERIALS IN THE 1950s

Bill Muir of Grand Rapids Label remembered the pressure sensitive materials of the middle 1950s:

> *For the first few years the materials that Kleen-Stik made were pretty good. They had, as I recall, a 20 which was a removable adhesive, a 40 that was semi-permanent and an adhesive designated 60 which was their permanent adhesive. Fasson then came out, they really had S201, which was their permanent adhesive. They had R100, which was their removable adhesive. And both...were extremely difficult to strip, if not impossible. [By applying heat to strip the matrix] we got some pretty good production out of the New Era press....It would never run through a Mark Andy....I remember with Kleen-Stik there were times when we were shipping more material back than we were keeping because the release was so inconsistent.*[82]

Ed Mahnic, Fasson's third salesman, talked about the materials as well:

> *Well, we had 50 pound backing...and we had 40 pound later in rolls, we had 80 pound, we had the vinyl. We started out with white vinyl and yellow vinyl. That was our first sheet stock.*[83]

Mahnic also found problems with the materials, such as an easy-release on vinyl that wouldn't release the backing at all.

> *So [the account] said "Okay. Show me how to get the backing off." So Howard, who was almost my size—six foot three, 240 pounds—he and I were holding this material...and we're pulling with all our might and we can barely move it....We couldn't pull the damn backing off.*[84]

In contrast, Ed remembered an express company in Akron, Ohio, which put big Indian heads on the trucks with a permanent pressure sensitive adhesive. "About eight trucks left and after about a quarter of a mile there were these [Indian head]...signs [that] fell off the trucks. I had some real problems, some traumatic things."[85]

A ROTARY PERSON

One of the amazing ironies of the pressure sensitive adhesive roll label business is that the two men who did the most to create the industry never met. Mark Andrews, Jr., says to the best of his knowledge, his father never met Stan Avery:

> *It really is amazing. I guess Dad's concept was to build a little machine and get those little machines out and have [both] a converting business and a machine business. Stan's concept was, I think,... after he invented the first [pressure sensitive] label was to build on that and build an organization, which he did just as planned. So they really took two different approaches...and never really met each other down through the years.*[86]

Andrews, Jr., believes his father did meet Burt Morgan in the early 1950s and could see the possibilities of a printed label. But this would require a shape—circle, square or rectangle—not simply a piece of tape torn off at a perforation:

There were methods for die cutting on a flat bed, but my father had no interest in flat bed. He was a rotary person from the very beginning. He didn't really care for reciprocal action, everything he wanted to do was rotary. So he decided to develop a rotary die.[87]

Andrews' die cutting roll was ordered from General Trademark in New York. Lee Carlson of Preston Engraving owns the letter, dated April 15, 1955, in which Andrews orders three rolls engraved:

Dear Mr. Capuzzo: Under separate cover I'm sending you three rolls to engrave. It seems that we are going to have lots of these, for there seems to be much interest in this if we can get it going just right. I attach the instructions for the three rolls. I have also attached some literature on our presses....One of these presses is on display at Gibbs Brower [Mark Andy's New York agent]...in your city. Now I offer you one of our presses in exchange for your engraving work. What do you think? Kindest regards. Cordially, Mark Andrews.[88]

THOSE DIE MAKERS

"One day a gentleman came in and said, 'I need a rotary die. And I understand you engrave in the round.' That started it. That's when I made my first rotary die."[89] Jim Ferris, who was with Preston Engraving for about forty years, remembers well that first order, sent to No-Lik Products of Maynard, Massachusetts. Ferris also remembers that the first die fit and worked immediately.

He gave me the proper dimensions for the die size, however it was up to me....See, the secret to making a rotary die is in the relationship of the cutting edge to the bearer so that the material at the cutting edge only cuts through the face of the material but does not damage the liner....That was what I immediately came on to. That was the start of our success.... First we always wanted a sample of the material so that we could proof the rotary tooling before we sent it out to the customer....We had a goodly number of presses... so we could test a die in a piece of equipment similar to theirs.[90]

Ferris believes that Preston Engraving was the first in the U.S., possibly the third in the whole world, to make an engraved rotary die. Tony Monaco, founder of Preston Engraving, a mechanical engraving company specializing in the aircraft industry, had hired Ferris to do some of the engraving. From this order came the beginning of the rotary die business for Preston Engraving.

ROTO-DIE

Richard R. Rosemann was a machinist working for Mark Andrews, Sr., in the early 1950s. In a magazine article he says:

> *The evolution of the Mark Andy press, from a tape press to a tape and label press, was quite fast. Early in 1954 I saw label stock for the first time. A roll of it was handed to me with instructions to make a cross-cutting roll that would cut only the top layer. I made up a standard cross-cutting roll, and added bearers for controlling the depth to which the inserted knife blades could cut. When I demonstrated the results to Mark he was elated....[neither] I nor any of the others present at the time, had any idea as to what was so great about a one inch wide strip of material with cross cuts 3/4 inch spacing.*[91]

Rosemann goes on to say the first die engraved for Andrews was made by a St. Louis packaging company, Charles K. Sweitzer Company, who wanted nothing to do with rotary dies. Then Rosemann made the die blanks sent to General Trademark, who engraved the second die "early in 1955. Before the year was out a Chicago company [name not recalled] was doing the engraving."[92]

Andrews suggested Rosemann learn how to engrave so he could supply the dies needed. "I engraved my first die in June of 1956. Roto-Die was incorporated March 4, 1957."[93]

YOU COULD EAT YOUR MISTAKES

Dick Capuzzo of General Trademark in New York recalled the business his immigrant grandfather had founded. In Italy the grandfather's first job had been making printing rolls. Out of what? Molasses.

"And it was great," Capuzzo said, "because it was one of the jobs where you could actually eat your mistakes."[94]

On arrival in New York Capuzzo's grandfather started work as a pressman for the printer who printed the Sears & Roebuck Catalog. In the late 1920s, he opened his own business and one of his earliest jobs was producing labels for bottles of bootleg liquor!

Dick's father joined the firm in the 1930s and was basically running the business at the age of eighteen. During World War II he worked a full day as a chemist, then a full day at the printing plant. "One of the things he got into in the late 1940s and early 1950s was aniline printing"[95](later to be known as flexographic printing).

They also were printing gravure and the war forced them to build their own cylinders because they could no longer be purchased from France. This led the company to begin producing hot iron transfers, which in turn led them into producing labels:

> *The stockings and socks, all the hosiery products, were held together with a label that was called a rider. What it was, was a gummed label....The labels would be wet and stuck onto the socks which would hold them together. They are doing it now [with] pressure sensitive.*[96]

Capuzzo believes they were in the hosiery business before the war, because they were using a seal press purchased from Keisey Press of Germany (now Keis and Gurlock).

After the war they obtained some pressure sensitive material and put it on the seal press, and were able to die cut it with flat dies. But the ability of General Trademark to make the printing cylinders led to their being contacted by Mark Andrews. Capuzzo remembers this as occurring in 1956:

> *And we manufactured a die, in fact a couple of them I believe, but one went to Rich Rosemann who worked for them [Mark Andy] at the time, and who now, of course, is Roto-Die. And Rich worked the die. In other words, we did the preliminary stuff because we had this pantagraph machine and we could take a flat design and make it into a rotary die....Then he touched it up a little, ground it here and there, did some work on it and was able to get it to die cut labels.*[97]

Both the Andrews, Senior and Junior, remember that it was the elder Andrews who sharpened the dies to make them work.

As mentioned earlier ("Competitors, Entrepreneurs, Innovators"), during the second World War Avery made a rotary die for a big U.S. Navy order, but it wasn't put into service at that time, and the official history makes no other reference to rotary dies. Yet it is most likely they did develop a rotary die before the mid-1950s. By then Avery was utilizing New Era letter presses (as were many other label and tag manufacturers), but each was modified extensively by Avery before being placed into service.

Andrews, Sr., in his 1978 tape recording, believed that he could "claim to be the originator of the rotary die cutting of label stock."[98] As happened a number of times in the industry, more than one person invented a product or a process independently at roughly the same time. For example, the Top Flight Corporation of York, Pennsylvania, produced a flexographic printing press about the same time as the Mark Andy press was first produced. And there was a Chicago company that produced rotary die cutting rolls at almost the same time as did General Trademark.

The important thing for the industry as a whole was that shortly after the patent suit had been decided, a person could buy a Mark Andy label press for $1000 to $1600. If one had a basement or a garage and two to five thousand dollars, one could get into the business.

ENTREPRENEURS WANTED—$8.00 AN HOUR

When Andrews began to sell his tape printers he also began to sell the concept of getting into your own business. He described the market opportunity, the endless actual and potential applications and the cost and returns of the business. In 1952 he was actively pushing machine sales and encouraging people with entrepreneurial drive and sales skills, which he felt was the key to success in the pressure sensitive tape business. He stated that you could make $8.00 or more an hour on a piece of equipment that cost you less than $1,000. With one or two decent customers, you could recover the investment in one to three months, using your wife as a press operator. Says Mark Andrews, Jr., "One of my father's really interesting ways of dealing with this industry was to drive everywhere in the country with a flexo machine in the back of the station wagon. If he couldn't sell it he would give it away and tell the person to pay for it when they got around to it."[99]

The early Mark Andy's were not exactly works of art. For one thing, Andrews, Sr., never made the same machine twice. Sometimes the customer asked for a special feature, sometimes Andrews just saw an improvement he could make, based on his last machine.

The web was tiny, two inches. When Dick Capuzzo's grandfather first saw one at Gibbs Brower, he called it a "piece of junk."[100] However, all early flexographic printing could not compete with letterpress, offset, or gravure in terms of quality. Everyone considered it to be the "junk end of the business."

Rubber plates would expand or shrink when the pressman least expected it. Flexography was a four season business. What worked in summer was no good in winter. The miracle was that so much acceptable work was produced at all.

But what other business could an enterprising entrepreneur commence for $2,000?

CHRONOLOGY OF APPLICATIONS, 1945-1955

1952	PSA Roll Label Patent Restriction Lifted
1954	Food Product-Primary Labels
	Variable Data Scale Labels
	EDP Address Labels
	Film Labels to Replace Decals
	Decorative/Novelty Labels
	Meat Package/Header Labels

CHRONOLOGY OF INVENTIONS/ INNOVATIONS, 1945-1955

1946	Mark Andy Flexographic Tape Printer
1950	Roll Label Dispensers
1951	Permanent Pressure Sensitive Adhesives
1952	First Silicone Coated Release Liner
1954	Mark Andy PSA Label Press
1954	Rotary Dies
1955	Improved Anilox Rolls/Ink Control
	Silicone Coated Densified Kraft Release Liner
	Automatic Label Dispensers

THE MAN WHO WOULDN'T QUIT, PART TWO

When Harold Scherer came to work at Kleen-Stik in March of 1948 he found they were bringing the web across the room over hot plates with fans behind them to exhaust the fumes out of the window. They were also manufacturing their own adhesive with a drill press and a wooden paddle. It must have been something of a shock:

We were on the south side of Chicago on the eighth floor, I believe it was, and then we took over another floor and another floor....It was my job to find a plant or have one built. We built it up in Harlem and Irving Park Boulevard. [It was the] Chicago plant at the time....We started out with one story because of the [building] code that was in Harlem, Illinois....A number of years later they allowed a second story.[101]

[1]Reminiscences of Mark Andrews, Sr., recorded May 28, 1978, p. 2.

[2]*Ibid.*

[3]*Ibid.*, p. 3.

[4]*Ibid.*, p. 4.

[5]*Ibid.*, p. 7.

[6]*Ibid.*

[7]*Ibid.*, p. 8.

[8]*Ibid.*

[9]John Garber interview with Bill Klein, p. 1.

[10]*Ibid.*, p. 1, 2.

[11]Clark. The First Fifty Years, p. 46.

[12]*Ibid.*, p. 50.

[13]*Ibid.*

[14]*Ibid.*, p. 51.

[15]*Ibid.*, p. 53.

[16]Communication with Ferdinand Rüesch, Jr., March, 1994.

[17]Wayne Lasinsky interview with Bill Klein, p. 11.

[18]John Orlando interview with Bill Klein, p. 19.

[19]*Ibid.*, p. 22.

[20]*Ibid.*

[21]Dick Capuzzo interview with Bill Klein, p. 13.

[22]*Ibid.*, p. 13, 14.

[23]Orlando interview, p. 10.

[24]*Ibid.*

[25]Orlando interview, p. 4.

[26]Art Schwartz interview with Bill Klein, p. 4.

[27]Lasinski interview, p. 1.

[28]Clark. The First Fifty Years, p. 57.

[29]*Ibid.,* p.61.

[30]Harry Stillman interview with Bill Klein p. 3.

[31]*Ibid.*

[32]Herb Smith interview with Bill Klein, p. 11.

[33]*Ibid.,* p. 2.

[34]*Ibid.*, p. 5.

[35]This is Harry Stillman's recollection. Others—George Collons and John Orlando—say uncle and nephew.

[36]George Collons interview with Bill Klein, p. 8, 9.

[37]*Ibid.,* p. 9.

[38]Clark. The First Fifty Years, p. 61.

[39]*Ibid.* The Thanksgiving date meant that the case was heard in 1951 and decided the following May 16, 1952.

[40]*Ibid.*

[41]Collons interview, p. 9.

[42]"Label Industry Honors 'Father' of Self- Adhesives." Paper, Film & Foil Converter, Jan. 1993, p. 78.

[43]Jim Kyte interview with Bill Klein, p. 2.

[44]*Ibid.*, p. 3.

[45]*Ibid.*

[46]*Ibid.*, p. 4.

[47]*Ibid.*, p. 5.

[48]*Ibid.*, p. 7.

[49]*Ibid.*, p. 10.

[50]*Ibid.*, p. 12.

[51]Mark Andrews, Sr., reminiscences, p. 8, 9.

[52]Bob Irwin letter to Bill Klein.

[53]*Ibid.*

[54]*Ibid.*

[55]*Ibid.*

[56]Walt Webster interview with Bill Klein, p. 1.

[57]*Ibid.*, p. 2.

[58]*Ibid.*, p. 3.

[59]Harold Sidrane interview with Bill Klein, p. 3.

[60]*Ibid.*

[61]*Ibid.*, p. 4.

[62]Al Sobel interview with Bill Klein, p. 3.

[63]*Ibid.*, p. 3.

[64]*Ibid.*, p. 4.

[65]*Ibid.*

[66]Bill Eiseman interview with Bill Klein, p. 5.

[67]*Ibid.*, p. 6.

[68]Bob Klas interview with Bill Klein, p. 2. It was probably the second Mark Andy sold as Andrews made a number of presses for his own conversion business.

[69]*Ibid.*, p. 3.

[70]*Ibid.*

[71]*Ibid.*, p. 21.

[72]*Ibid.*, p. 22.

[73]*Ibid.*

[74]Dale Bunnell interview with Bill Klein, p. 8.

[75]*Ibid.*

[76]Mark Andrews, Jr., interview with Bill Klein, p. 6.

[77]*Ibid.*, p. 7.

[78]Paul Styers interview with Bill Klein, p. 6.

[79]*Ibid.*, p. 8.

[80]*Ibid.*

[81]*Ibid.*, p. 9.

[82]Bill Muir interview with Bill Klein, p. 10.

[83]Ed Mahnic interview with Bill Klein, p. 6.

[84]*Ibid.*, p. 14.

[85]*Ibid.*, p. 16.

[86]Andrews, Jr. interview, p. 8.

[87]*Ibid.*, p. 9.

[88]Lee Carlson interview with Bill Klein, p. 4, 5.

[89]Jim Ferris interview with Bill Klein, p. 1.

[90]*Ibid.*, p. 3, 4.

[91]Copy of magazine article, "Rotary Diecutting Pressure Sensitive Labels—How it Started." Sent to Bill Klein by Rosemann, January, 1993.

[92]*Ibid.*

[93]*Ibid.*

[94]Capuzzo interview, p. 3.

[95]*Ibid.*, p. 5.

[96]*Ibid.*, p. 7, 8.

[97]*Ibid.*, p. 9, 10.

[98]Andrews, Sr., reminiscences, p. 9.

[99]Mark Andrews, Jr. interview, p. 3.

[100]Capuzzo interview, p. 11.

[101]Scherer interview, p. 9.

SECTION THREE
ALL THE PIECES BEGIN FALLING INTO PLACE:
1955-1960

THE GIANT MOVES

Since Avery Adhesives could not stop others from producing pressure sensitive labels, the company decided to create a separate division to supply materials to other manufacturers. This decision was made at about the same time Avery realized the company needed to create manufacturing operations beyond the southern California area, and should begin to look into the international market.

Each of these three decisions meant a major transformation in company operations. That each succeeded is a tribute to astute Avery management in the early to mid 1950s.

INTERNATIONAL AVERY

Avery had made the decision to operate a converting business and a base materials operation in the United States. In the mid-1950s the company took the same tack in Europe.

A base materials plant was opened southwest of Amsterdam in Leiden, Netherlands, in 1955. It was fully operational in 1956.[1] Also in 1955, The Antonson Company of Sweden became an Avery Label franchise. The second franchise went to William Sessions, Limited, of York, England.

Some other roll label manufacturing licensees included:

VRG Paper	Benelux, France
Zweekform Werk GmH	Germany, Austria
Heusser	Switzerland, Italy

The decision to sell to franchise competitors would cause trouble in the 1960s.

FASSON BECOMES AVERY'S COMPETITION

The first sales of Avery base materials came, logically enough, from the Monrovia plant. Avery had an important product difference that gave the company a competitive edge over Simon Adhesives and Kleen-Stik. It had a release coating that permitted assembly line label application.

So, in 1954 the decision was announced that its new materials division would "sell base stock directly to printers, converters and label manufacturers, many of whom will be selling in direct competition to the label company."[2] This created some real problems within the organization. Russ Smith recalled:

> *The demand for our base materials from competitors posed a real dilemma for us. How could we sell to those companies and compete for their customers at the same time?*
>
> *We were determined that we were going to handle the matter of selling to our own competitors in the right way, so that people would know that they could depend on our integrity. Potential customers for Fasson had to know that they could talk to the Fasson representative with complete confidence that the information would not flow back to Avery Label. There were also some who feared that we would rig the prices, so that the competing label companies could not be competitive.*
>
> *Our answer finally was to set up Fasson as a completely separate division to manufacture base materials for the entire label industry. We kept our own Avery Label people out of the new materials operation, and we kept the base materials people out of the Avery Label plants. We wouldn't even let an Avery Label salesman in a Fasson plant or vice versa.*[3]

In 1953 Stan Avery met Burt Morgan, a man with considerable experience in manufacturing medical and masking tapes. He offered to locate, build and run a PSA coated materials plant, and by February, 1954, the 12,500 square-foot facility in Painesville, Ohio, was opened. A year later it was more than doubled in size.

Burton D. Morgan is one of the more colorful men to have worked in the pressure sensitive business. He graduated from Purdue University in 1938 with a BS in mechanical engineering (he received an honorary PhD from Purdue in 1992). He began working for B.F. Goodrich, ending up as chief engineer for the company's synthetic rubber plant in Louisville. From 1943 to 1946 he worked as a mechanical engineer for Johnson & Johnson, the makers of Band Aid bandages. From there he was hired by Bear Manning specifically to help them move into the pressure sensitive field.

Feeling that he was vastly underpaid at $9,500 a year, Morgan called Avery Adhesives and asked to speak with Stan Avery. Morgan told Avery he wanted to make plain label paper. "So he [Avery] said, 'Well, why don't you come out and talk to me?' 'Well, I can't afford to.' So he [Stan Avery] paid my airfare to California and we went and formed a company called the Avery Paper Company...in Loudonville, New York."[4]

But, for manufacturing, Morgan chose Painesville, Ohio. Why?

> *I had to locate near the research centers for polymers. That was Akron. I didn't think too much of Akron, but this was close by on the lake in a nice community looking for new business.*[5]

To construct the new Painesville plant, Morgan dealt with D.D. Davis, whose business was to build plants—at his own risk—for new companies.

> *All you had to do was sign a lease and he would buy the land and build the plant....So D.D. Davis came up to Painesville and...he went to the city fathers and said, "Okay, that piece of land you've got, I'm going to build a building for this fellow here, and I've got to have titles to the land. How much do you want for it? [They said $120,000.] He wrote them a check and said, "I own the land, right?" "Well, it will take a week to write the deed and it has to go through city council." "Yeah, well, I'm going to start building tomorrow. If you want to stop me you have to give me the check back and I'll go home"...They let him do it.*[6]

Morgan recalled that Davis asked who the largest concrete dealer was, then called the company with an order for five cubic yards every 20 minutes, starting at 10 A.M. At 7 A.M. the next day the surveyors came and asked Morgan where he wanted the building. He told them. They put one stake in the ground and laid out the building. At 8 A.M. a big back hoe trailer arrived and the back hoe started digging an 18 inch wide trench three feet deep. At 9 A.M. a truckload of steel arrived and at 10 A.M. the first of the concrete trucks appeared.[7]

A MARX BROTHERS MOVIE

The first few months of Fasson were nothing if not exciting. William B. Dodge, Fasson's first plant manager (later marketing manager, then general manager) referred to himself and the others who began operation in 1954 as amateurs:

> *We had people trying to run this machinery who had no experience with anything like it. Some of the guys we hired were farmers and some of them were boxcar unloaders from the freight yards. They didn't know which end of the machine was which. We were all completely bewildered.*[8]

Morgan remembered the first product, a one-sided coated pressure sensitive paper for covering stainless steel. "And to this day we've never sold any of it."[9]

The First Fifty Years says the first big order, with production starting in November, 1954, was for General Tire and Rubber. The product was Decor-Ease, a self-adhering shelf liner. Bill Dodge says it was extremely difficult to make:

> *In the beginning, we were wrecking the orders all the time. What made it so difficult to get running was that vinyl film moves and stretches, especially when you put it in a hot oven. When vinyl gets warm and sticky it pulls out like a big rubber band.*[10]

Both the official Avery history and Burt Morgan agree the breakthrough order came from Kimberly Clark. Avery says it was for shelving paper;[11] Morgan says it was for Marvelon, a pressure sensitive wallpaper.[12]

No matter what the product, it necessitated a much larger plant and a sixty-inch coater—twice the width of the one machine they had.

Desperate for the order, Morgan said he would have everything in place within 90 days. (The machine was 60 inches wide to accommodate a 54 inch web, which could then be slit into three 18-inch wide rolls.)

Morgan took the drawings from the mechanical engineer for the existing coaters, laminator and ovens, and added thirty inches to the middle of each.[13] The drive itself would prove even more difficult. Morgan bought drives from Reliance Electric. The salesman was Ron Bigger, who later became president of Nestlé, U.S.A. Bigger told Morgan he could produce a drive in one year. Morgan had him take the order anyway. Then:

A couple of days later I called up Edward Helm, president, [and I told him] I have never even gotten quotations from anybody else. Just Reliance....I'm a 100% Reliance man and I've never asked you a favor in my life. And I'm getting ready to ask you a big one.[14]

Helm agreed. In 45 days the drive arrived. Morgan says that years later he asked Helm how he did it and was told, "Well, we have a division here called the breakdown division. When an important company has a big breakdown we bring the stuff in here and fix it, we work 24 hours. All I did was take your order and hand it to the breakdown division."[15]

"WE DESTROYED OUR OWN BUSINESS"[16]

One of the early products Avery Paper made was the material for bumper stickers. According to The First Fifty Years, the bumper sticker was a creation of Gill Studios in Kansas City, Kansas.[17] It was designed to be a premium product so the buyer could easily remove the sticker after a month, or even a year, and leave no mark.

Kleen-Stik had most of the market. To make inroads, Fasson lowered the price. Kleen-Stik followed. Fasson lowered the quality and the price, and Kleen-Stik followed. The bumper stickers went on easily enough now, but they would not come off. Burt Morgan notes:

So not only did we destroy the market... pretty soon nobody bought them and they were all attempting to sue us....We'd just laugh at the lawsuits. "Go ahead, if you want to sue me, I'll give you the key to the place." Gerry [Cole of Kleen-Stik] and I talked later in life [about the bumper sticker wars, and agreed] we should take 30 minutes and just stand here and kick each other. We destroyed our own business.[18]

HAZARDOUS TO YOUR HEALTH

Fasson was using solvents to make adhesives in much larger quantities than at Monrovia. "Those fumes," remembered Bill Dodge, "were intoxicating. So you got a pretty good high working at Fasson."[19]

Some guys would hear a radio playing when there was no radio. They got all banged up like these kids do today sniffing glue. Because that's exactly what they were doing. We would have to take them outside to get their heads cleared.[20]

The fumes were bad enough. The fires, often caused by static electricity, were much worse. Bill Dodge recalled:

The fires usually started in the churns, where we mixed the rubber with solvents. First we would take big blocks of synthetic rubber and grind them down to the size of pellets. It took a great amount of power to grind up the heavy rubber, and in the process the rubber would acquire a tremendous static electric charge.

We would carry this rubber, with a big static charge on it, over to the churn. The churn had maybe 800 gallons of hexane solvent in it, with a layer of fumes all over the top. It also had a static charge of opposite polarity from that in the rubber. We'd open the lid of the churn, dump the rubber in, and WHAM, we had a fire.

The top of the churn would go flying and the flames would shoot up. It was pretty frightening.[21]

A FIRE TRUCK A DAY

Chuck Reed, who joined Fasson in 1957 as a chemist, recalled the almost daily visit of the Painesville fire trucks:

I'm glad they didn't think we were crying wolf and not show up....[Our] guys learned how to put them [the fires] out in a hurry. [But] we only brought out adhesive and lacquer one drum at a time. We were aware we could have a potential problem.[22]

Reed also remembered that even his own company wasn't especially helpful in the early days:

[I was] on the road keeping products sold, in other words, teaching people how to convert it so they could sell it again, [and] the guys back at the ranch would make product changes, not tell me about it and then I was in the field wondering what was going on...for instance a liner a quarter mil fatter than it had been before and all of a sudden the tooling that existed in the field couldn't handle the stock.[23]

Since he was the only technical/service representative, Chuck Reed knew many of Fasson's earliest customers:

Don McDaniel was always one of those guys with an open mind who wanted to try stuff. [Steigerwald] was always good to go in on test material. If you had something new and different they would say, "Hey, think about us." They liked us to come in and work with them....Finally, it got so bad I was spending a day a week in Chicago. Mostly again, Steigerwald.[24]

Reed notes that Giles Clark (Fasson's Chicago salesman at the time) was a less than perfect driver:

He got tickets all the time. It got to the point that when he picked me up at the airport, I'd automatically say, "Give me the keys"...I hated being in the car with him.[25]

THE PRICE IS RIGHT

The 1950s brought a dramatic change in the price of pressure sensitive material. In 1952 the price per thousand square inches was about 27 cents. By 1954, when Fasson was opening its plant, the price dropped to 20 cents. Further competition dropped it still farther, to about 13 cents.

Many people interviewed for this book said that Morgan Adhesives (MACtac) lowered prices when it opened its doors, but Morgan himself denies it.

That's the common perception and it's not true. When we started at Morgan, Fasson knew I knew all their customers. Some would buy just from friendship. So they passed around the word that the way Morgan is going to get into the business, they're going to cut the prices. This was the best thing that ever happened. Everybody called me up for the new low prices. They didn't get the low price; they just thought they did.[26]

Al Norman of MACtac sums it up well, "He brought a paper mill mentality to pressure sensitive converting business. It forced everybody else, as difficult as it was to adapt, to rethink where they were and what they were doing."[27]

-007-

Another early Fasson employee was Bill Donovan, who joined the company (probably) in 1956.[28] He became the seventh salesman hired—his sales number was 007—and worked the New England territory.

Back in the early days, of course, pressure sensitive was not exactly a great product. It had all kinds of quality control problems....The product was not very stable. The release liner was a product Stan Avery had pretty much developed himself. [Then] it was a parchment type, what we called the old firm release. It was okay for Avery [Label] because everything they did was designed and built to handle that product. But the few converters we had in the industry...were working primarily with lower cost flexographic rotary presses. Our product at that time just didn't work.[29]

Donovan points out that the traditional tag and label people used tag and ticket presses which ran slowly and could be adapted to convert almost anything:

But the product that was really coming along in volume was the flexographic machine, considerably lower in cost to buy, considerably greater output from the equipment. In that first year and a half, Fasson was at a real disadvantage.[30]

The competition was not Kleen-Stik, but Simon Adhesives Products (Sander Simon) and Pressure Stick (Mark Simon). "Simon [Adhesive] and Pressure Stick [were] at least temporarily superior to everything Avery [Fasson] offered."[31]

RELEASE LINERS

Most of the earlier release liners were of glassine, a super calendered, smooth dense transparent or translucent paper with a high degree of hydration. Glassine was often used as an envelope window. Its highly uniform caliper proved useful as a release lining, but it was exceptionally brittle and therefore hard to use, especially at low humidities. Deerfield Paper sold primarily to Fasson, while Westfield River supplied Kleen-Stik.[32]

> *In the mid-fifties Crocker Burbank of Fitchburg, Massachusetts, began the production of a supercalendered kraft paper (SCK). About 1960 large quantities of EDP labels were printed on New Era presses and fan folded with perforations added. The "burst perf" was impossible with a glassine liner because glassine was too brittle, but worked perfectly with SCK.*[33]

The new generation of release liners are not papers at all, but plastic film, expensive when compared to SCK, but more dependable on high speed automatic labeling lines. They also have the added advantage of ease of recycling. Mobil Chemical's Label Mate™ and several other entries are a more recent lower cost plastic alternative to the polyester type.[34]

FASSON FANTASIA

Burt Morgan had hired twelve salesmen by 1957 and a year later it was clear there was a revolution going on in the industry. While sales of industrial type pressure sensitive adhesive coated products actually increased, its share of Fasson output dropped from 85 to 15 percent. Sales to the graphic arts industry accounted for all the rest. Instead of a few customers (mostly General Tire and Kimberly Clark), they had dozens and dozens.

Fasson salesmen sometimes would drum up pressure sensitive orders and hand it to a printer, hoping he would buy their sheets or rolls to fill the order.

In addition to direct salesmen, Fasson sold through distributors. According to Don Buchta, his father's Textile Craft Products was:

> *...actually the first distributor that Fasson took on....in fact, we were their first distributor that attended their sales meetings. We not only took on the sheet material we also took on the roll material as well. Knowing that this product could be run on flat bed reciprocating presses that they use in the tag industry, we called on the tag people to show them how they could run pressure sensitive material on their equipment.*[35]

There were other, less productive ideas as well. Bill Dodge remembered:

> *When we were scrambling around in the beginning, trying to come up with new uses for pressure sensitive materials, we tried lots of dumb ideas....I think the worst was Burt's idea for pressure sensitive paint. Instead of painting your house, you would stick on strips of Fasson white vinyl.*
>
> *One summer Burt had two kids paint his house...with this vinyl.*[36] *By next spring, the stuff was peeling off and blowing around everywhere.*[37]

PLAIN-VU LADIES

Burt Morgan's book, Start At the Top, contains an interesting entrepreneurial anecdote he calls "Plain-Vu Ladies."

Two housewives had concocted a transparent adhesive that enabled documents such as birth certificates and drivers' licenses to be permanently covered in clear plastic. Morgan tested their adhesive for them, then offered to make a run of Plain-Vu, including plastic, adhesive and backing paper, cut to their specifications. He charged them $500, which they paid in small bills.

Two months later the ladies returned to order more. The ladies were selling by direct mail and already getting repeat orders.

Soon thereafter two men came to the office. They were in the direct mail business and had purchased the Plain-Vu business. "These two ladies demanded cash and we gave them $75,000 for everything—the trade names, the remaining inventory and the sources of supply."[38] In four months the two entrepreneurial women had made over $35,000 each.

STAR WARS

According to The First Fifty Years, Morgan had been given a 20 percent share of Avery Paper Company (later to be called Fasson) and, in 1958, Morgan brought in a group of investors to see the Painesville plant and to put up the money to purchase the remainder of the company. As reported by several Fasson executives, Morgan also held a "secret meeting of Fasson's top officers to reveal his plan and to ask them to join him."[39]

Morgan remembered it differently. He said:

> *Well, after Fasson got going, boy, it was making money like mad and Avery Label Company...was not doing too well. What we had done was put every printer in the label business. And he was suffering so he wanted to buy me out. And I wouldn't sell....[Stan said] "even though we're friends, I'm going to have to fire you....If you agree never to go into the pressure sensitive business again...I'll give you $400,000 U.S. dollars. If you won't sign that I'll buy them [your shares] from you for $160,000." I don't know were he got those numbers.*[40]

Morgan states he took the $160,000, but it wasn't paid in cash; the money was to come over three years. With the help of the Bemis Bag Company's Gregg Bemis, Sr. (whom Morgan calls "the wise and little old Vermonter"[41]), Morgan was able to start up a competing firm in Stow, Ohio, forty miles south of Painesville. In his book Start At the Top Morgan describes why Bemis decided to back him. Morgan had described a three-year plan of projected growth. Bemis asked about his twenty-year plan.

> *With that I took a piece of graph paper and while he watched I wrote the sales volume figures up the left hand margin, going from 0 to 40 million. Across the bottom margin I wrote the years from 1959 to 1979. Then while he watched I drew a curve, concave upwards, from the lower left-hand corner to the upper right-hand corner. The chart later became famous because we were right on the line for the first five years, after which we did much better than planned.*[42]

Morgan had great confidence his company would succeed. The first thing he did was to get a letterhead printed, which read, "Morgan Adhesive Company, World's Greatest Producer of Adhesive Products. We'd never made any."[43]

Bill Donovan notes that "Morgan believed that the industry should not be dominated by one company. [He] broke with Avery, started a competitive company and succeeded in opening up the entire industry, greatly increasing total usage."[44]

THE OHIO YEAR

To make a concrete show of support for Fasson, Stan and Dorothy Avery moved to Painesville, Ohio, in early 1959, and took personal charge of that operation. Avery said he enjoyed his year in Ohio, "It's the nearest I came to really running something myself after the very early days."[45] He even joined a bowling league and rolled a five in one game! In August of 1959, Avery chose William R. Zimmerman, a director of Moraine Paper Company and American Envelope Company of West Carrollton, Ohio, to be general manager. Bill Dodge became marketing director and Les Dickard, research director.

ALL OR NOTHING

Herb Smith remembered Southwest Tape and Label and Contact Products:

3M had a small tape printer in the lab, which was my first introduction to printed tape. In sales, 3M had an incentive plan to promote printed tape sales and this got me interested. I resigned 3M Company, bought a used three inch Mark Andy tape printer for $600 and got started in 1956...as Southwest Tape and Label. The machine needed a lot of repair and necessitated many visits to the machine shop. It became obvious I could get the machine shop to build a printer better than what I had, so I did, and never bought a machine thereafter.

Our first big order...amounted to $2,300 for printed acetate film tape....Label stock came into being about 1957 and we started score cutting and die cutting and stripping. Sales grew from $40,000 in 1956 to $7.5 million in 1976. Employees grew from 2 to a maximum of 165.

We felt we had a distinct advantage over most competitors in machinery technology as we made our machines heavier and stronger than those offered on the market. We added new innovations immediately and were able to make special equipment for special jobs in minimum time.

Once we had a requirement to furnish 200 foot rolls of painted Tyvek marked off from zero to 200 feet for measuring distances in a Ford Motor Company "pass, punt and kick" program. We line hole punched 6 mil mylar and made a flexo plate 12 inches by 200 feet long, fed it into the printing head on sprockets and festooned the plate from the ceiling to the floor about ten times. Then we slit it into rolls 3 inches wide. The order was completed within two weeks, including the manufacture of equipment to do the job.

The company name was changed to Contact Products in 1967....The future looked bright. [In 1977] a company by the name of Dixico expressed an interest in buying, but I was not interested. They persisted and I finally gave them a figure that I thought was ridiculous and they accepted. I sold it. As so often happens they mismanaged the company and went Chapter 7 in 1988.[46]

POISED FOR GROWTH

About 1961 Avery commissioned a $50,000 to $60,000 market research study to determine what the company should do to increase its business. The report concluded that Fasson should be "spun off" and Avery continue on its own as a label producer. Bill Klein, who was then marketing research manager of Fasson, saw the report and sent a hand-written memo to Bill Zimmerman stating that the consulting firm "didn't know what it was talking about."[47] The note finally found its way to the Avery head office, the report was junked, and Avery began to "push the true story that Fasson was not linked directly with the label division. They were linked almost like a holding company."[48]

Both Morgan Adhesives and Fasson began to market their products aggressively, which allowed the price of pressure sensitive labels to drop significantly, becoming much more competitive with gummed labels. (Even before Morgan Adhesives, Fasson had been forcing the prices lower.)

In addition to lower prices, another factor in making PSA labels more acceptable was the general introduction of the silicone release liner and the steady improvement in the quality and availability of an ever greater variety of pressure sensitive adhesive roll label stocks.

GRAND RAPIDS

Grand Rapids Label, like a number of other old line firms, was in labels before PSA. These companies all saw the advantages of pressure sensitive materials and became part of the new competition.

Bill Muir's Grand Rapids Label was originally chartered in 1884 as a general printer. Around the 1930s they were producing a variety of heat seal labels. They enjoyed a close relationship with Oliver Packaging, a packaging machine manufacturer:

We produced the labels and they produced all the packaging equipment. We sold that segment to them in 1941 and had a non-compete agreement for five years. About the time the war was over and the agreement expired, the first clue toward pressure sensitive came into the marketplace.[49]

So rather than return to making heat seal labels (they were making gum labels all along), they chose to investigate pressure sensitive.

> *The product was more expensive but when you consider the ease of application...it became a viable alternative to the water decal.*[50]

Muir relates that quality problems were extremely troublesome in the 1950s when he was working part-time while in school. Everything "was produced on letterpress equipment and the Young Engineering and Pierce presses."[51]

Since quality was hard to obtain with the materials available at the time, Grand Rapids sold mostly industrial labels and secondary labels. "It wasn't until a number of years later that it really became a viable alternative to a primary label."[52]

THE MYSTERIOUS DEBUT OF LABEL-AIRE

Automatic application opened up the volume market for pressure sensitive adhesive labels, at that time dominated by wet adhesives and gummed labels. Yet the many people interviewed for this book remember the introduction of the Label-Aire automatic labeler as occurring at differing times. Even Dick Riley of Label-Aire says only, "early or late fifties."[53]

George Collons says 1958 "would be a good pick."[54] But there had been an earlier fully-automatic Kleen-Stik applicator demonstrated at the Graphic Arts Show in Chicago at what Collons recalled as the fall of 1950. Collons said it was designed by a "guy named Goldberg."[55] The airblow technology, which is what Label-Aire began with, was a west coast invention according to Kleen-Stik's Art Schwartz.[56] Schwartz, who started at Kleen-Stik in 1950, says the Label-Aire was introduced in 1959.[57]

It was the advent of the automatic applicator that opened up the larger label applications to pressure sensitive adhesive labels. It made the endless variety of PSA labels available with low cost automatic application. At the time, the cost of the then standard automatic water-gum labeler was $10,000 to $20,000. Due to size and complexity, the water-gum labeler had to be designed into the production line. An automatic PSA roll label applicator cost only $2,000 to $3,000 and required no special piping. It could be retro-fitted into any production line and at virtually any point. An equally important factor was that there was no need for clean up after the line was shut down as was necessary for the water-moistened and direct glue labelers. Kleen-Stik's Label-Aire became the most successful of the applicators.

BLADES, NOT THE RAZOR

Muir remembered taking an early Label-Aire and a Tuck Labeler in his station wagon and demonstrating the advantages of automatic equipment. "We were just trying to promote it in hopes they would use it, use more labels and we would sell the labels...the razor/razor blade syndrome."[58]

HERE, THERE, EVERYWHERE

Britain and America were test flying jet aircraft at about the same time the German jet fighter appeared in the air to harass the U.S. Flying Fortress. Inventions seem to occur precisely whenever the need for such an invention exists.

In approximately 1957 a customer walked into Preston Engravers in Connecticut and asked for a rotary die. The company, then owned by Tony Monaco and Jim Ferris, made steel stamps and marking devices. The customer, No-Lik Products of Maynard, Massachusetts, had ordered a flexographic press made by a Milwaukee company called Frostead. Jim Ferris recalled:

He gave me the proper dimensions for the die size [but] it was up to me to see the secret of making a rotary die...the relationship of the cutting edge to the bearer so the material at the cutting edge only cuts through the face of the material but does not damage the liner.[59]

One of the Preston Company's techniques that led the company to impressive growth was that they would purchase a press similar to the customer's press so they could test the die before shipping it.

Jim Ferris believes his company was the first to make six, ten and 16-inch wide dies.[60] In a strange echo of Mrs. Avery's earlier direct mail success, a Preston customer asked Ferris if he wanted to grow. "Your product is excellent. You have no competition to speak of. Get out a flyer and get a mailing list." Which is exactly what he did. Ferris recalled there was only a half page of label producers listed for the entire country at that time. "A [label] stock manufacturer was kind enough to give me his mailing list. I put out the mailing and within three to four months, we had 30 new accounts."[61]

Turn-around time has certainly changed in the business. Ferris says they used to be proud of a week to ten-day delivery; currently it's three to four days. Competition has grown as well. "When I was still in the business [he retired about 1985] there were roughly three rotary die makers in the whole world. Now in the U.S. alone, I believe there are at least 25."[62]

Lee Carlson, who joined Preston in 1960, recalled the incredible secrecy of the industry's early years:

> *It was extremely secretive. We would never allow anyone in the plant at all. [In many instances this was justified.] If a customer comes to you...to produce something proprietary, it doesn't leave the building. It is...an unwritten law if one of your customers is having difficulties producing something and you know [another customer] is doing the same thing...you are really bound by honor not to say anything. That's the way it has always been....Basically [the die makers] all did the same thing....I mean we do things our way, Roto Metrics does things their way, Rollcut [does] their things their way. It used to be taboo to even talk to [a competitor]. Just this past year at Label Expo [1993], we took some of the people from Roto Metrics to lunch. Actually twenty years [earlier] I could have got fired for that.*[63]

For much of Preston Engraving's corporate history it was owned by Ferris, Martin Keston and Paul Monaco, son of the original partner. Since the trio sold out it has gone through three changes of ownership. It was purchased by Roto Metrics in late 1993, but many of the original employees remain.

THE SMUGGLER

Dick Egan, who now owns Seal Pack Corporation, got started in the tape business in 1948 at Mystic Tape of the Mid South in Louisville, Kentucky. Mystic made primarily industrial tapes and a decorative cloth tape sold to a broad range of retailers.

Mystic's largest customer was General Electric's Appliance Park. They were searching for printed tape and Egan says they were aware of Mark Andrews and his narrow web one color tape presses. But since Andrews had some tape business in Louisville, he would not sell Mystic a press. So...

Ted Hattemer (owner of Mystic Tape and Adhere Products) did find out there was a press in Canada...I think it was probably a three-inch, one-color press. No bigger than a suitcase. And he bought that thing...probably paid about five hundred or a thousand for it in 1957 or 1958. And he brought that back. He literally "snuck" across the border.[64]

Soon they had more orders than they could print, "At one time there was something like twenty different pieces of printed tape on each washing machine at Appliance Park down there....We just couldn't keep up with it."[65] Mystic Tape was sold and Adhere Products was incorporated to pursue the PSARL business.

In the late 1950s the Mark Andy Company began concentrating on machine sales (it sold the converter business in 1959), so Adhere added more presses. "I think...four inches probably was the widest press and maybe up to three colors was the maximum."[66] Adhere made an early sale of two Label-Aire applicators to the Claire Meat Packing Company (now Fisher Meat Packing Company) in Louisville. Egan recalled:

I can remember days when we threw away more label stock than we shipped. It was so bad...the Label-Aire applicating machine was probably ten years ahead of the pressure sensitive converting equipment and the material....It's basically a roll over a peel plate that is blown onto the product, but it was so demanding that the material that was being made and the converting equipment, primarily the dies, were really not up to producing a product that could successfully run through that equipment....I think Label-Aire forced the industry to improve itself much more rapidly.[67]

Egan remembered the problems with the dies very clearly:

The theory was they made it out of what we called a soft metal [therefore] you had to continuously work on the dies to keep them cutting. [We needed] a clean die cut so you could strip the matrix. I remember a little set of "tinging tools," we called them. They would be running and all of a sudden the matrix wasn't being cut clean. They would get up on the press and "ting" around on the cutting edge until they built it up so it would start cutting. And many times, of course, depending on the operator, they'd get too much and start cutting through the liner.[68]

Another problem was more chemical than mechanical,according to Egan.

[It was] the inconsistency of the release on the label stock itself. It would vary from one roll to the next—or right in the middle of the damn roll.[69]

GO WHERE THE COMPETITION ISN'T

Another 3M salesman who entered the business to print Scotch Tape was Lee Paul. Paul had been with 3M for a dozen years before starting his own business, Tape & Label Engineering, in 1958.

3M was and is one of the finest corporations in the country....[It] was a great training ground for me because without the help I received there in management and selling I could not have been successful in my own business...[but]I felt I would be happier working for myself.[70]

Paul had worked for 3M in the St. Paul home office, but decided to build his business in St. Petersburg, Florida, because he liked the city, "plus the fact that it was an area that had no competition" and "I wasn't competing with the people I had been working with while selling for 3M."[71]

One of the first applications Paul recalled was a primary label for Shakespeare Fishing Tackle, "but in those days primary labels were not as sophisticated as they are today."[72]

Interestingly enough, Paul believes he and Rich Rosemann of Roto-Die Company went into business on the exact same day. Tape & Label Engineering, now owned by Weber Marking, is one of the largest label producers in the southeast. It is a completely integrated plant except for ink, paper and dies. It has been a Label-Aire distributor for more than 35 years and was a pioneer in the use of UV (Ultra Violet) inks and coatings.

CHERRY PICKING

In the late 1950s Bill Swan, District Sales Manager of 3M in Cincinnati, Ohio, suggested Adhere move to Cincinnati. Cincinnati was a much bigger market and Swan said there were few, if any, converters in the area. Egan worked Monday through Friday in Cincinnati, then returned home to Louisville, before finally moving to Ohio in December of 1959.

Most of the major Cincinnati corporations, such as Proctor and Gamble, Krogers and Kahn's Meats, were using some pressure sensitive materials and they proved remarkably easy to sell:

> *All...the label stock was coming out of Avery Label in California....It was like picking cherries. You walked in there and said, "I'm a converter" and they almost jumped up and kissed me, because they were so [angry with] Avery. Every time I ran across Avery it was a sure bet I'd get this business. And we continued to add equipment and probably had to move every two or three years because of our growth. I guess we averaged over those years about 25 percent [growth] a year.*[73]

AND WE STARVED TO DEATH

Stan Fulweiler started in 1956 making pressure sensitive tape as a substitute for a 3M label end product. "Primarily made in laboratories. We tinted our own tape over there."[74] The material was made for Repro Templates, and when Fulweiler felt he was being treated unfairly he left to start making his own material for Repro Templates' Daisy printer:

> *And we starved to death. So we put an old Mark Andy in the attic of a house. All the money I could borrow was two grand. That would just about pay for one machine....That's how we started making industrial layout tape for Repro Templates. And I found out that was not going to support me or my family and finally I had to go sell something.*[75]

Using his engineering background, Fulweiler began to specialize. "We did things like electromagnetic shielding."[76] Fulweiler remembered working with label stock on a semi-wide paper which sometimes split and sometimes didn't.

This first company was called FD Tape Printing, Inc., but soon the company grew and in 1966 the name was changed to Tapecon. Fulweiler said that Tapecon developed over a great idea.

> *I figured out a method of putting sound on the back of photographs using a laminate and magnetic tape and double-faced adhesive tape....That was the invention that put voice on cards and paper and whatever else....But this one was just a hole in the ground to pour money into...was never able to sell it.*[77]

Fulweiler was just beginning his entreprenurial drive. He and his brother Raymond also started Wisconsin Label Incorporated, in Algoma, Wisconsin, in 1966.

As Fulweiler said, one thing led to another and soon there was a Peachtree City plant (outside Atlanta) called Voxcom Labels, which became a division of Tapecon. Later came the Buffalo plant and the Rochester plant.

> *The guys that were running the Rochester plant and the Buffalo plant bought us out with a leveraged buy out... because I was taking the profit and putting it into my inventions....So [in 1983] we ended up with Voxcom in Peachtree City. And Tom Hale and Harlow Davis ended up with Tapecon which they took private last year (1992).*[78]

Fulweiler still owns stock in Voxcom as well as Wisconsin Label, and all three companies were in the top 100 PSARL companies in 1992, he said.[79]

The companies specialized in different aspects of pressure sensitive. In Rochester, Tapecon was primarily technical, with customers like Kodak, Xerox, General Electric—"primarily converters, but the unusual thing was the straight label job."[80] Wisconsin Label specialized in high volume systems for food and cheese, but has now become a group of seven corporations, all in the Middle West. "We used to go to auctions of other label companies that folded. Go over with an old station wagon and pick up whatever pieces."[81]

Fulweiler summed it up well:

> *I've started other businesses, some of them have dropped by the wayside....Put together an outfit called Web Coating that ended up with the White Sewing Machine [Company]....You really have to develop your own business if you're going to do some of these things....Pressure sensitive label industry was a clean interesting business. The competition would help you if you needed...yet the following day you might be sitting across the table...trying to get the business away from him, but basically it was clean, and it was interesting, and it grew and it was very kind to the whole bunch of us.*[82]

EASY TO REMEMBER PRODUCT LINE

Ed Mahnic became the third salesman hired by Avery Paper, before the Fasson name change.[83] Leon Kaston, the first salesman, was based in New York, Jim McMann covered Chicago and Mahnic worked out of Cleveland. The fourth salesman was Paul Mailer in Detroit.

Their competition was Kleen-Stik, and originally everything was in sheets, although Mahnic recalled Kleen-Stik had some roll business on the east coast.[84] For a year Mahnic made nothing but cold calls on advertising agencies, corporate advertising departments and other potential corporate end users:

We had a brochure put together and a swatch book... with about eight stocks and that was it. White litho, tag, an orange-yellow fluorescent and two foils, silver and gold...[a total of] eight stocks, all sheets [with a] 100 and 201 adhesive. That was it.[85]

Later they added a C100 rayon acetate cloth. They offered 50 pound and 80 pound backings (40 came later in rolls) and white and yellow vinyl.

There were few converters in the middle 1950s. Mahnic recalled Salem Label, Kimball and York Label. Chicago had Steigerwald. Mahnic recalled an early carload sale of bumper sticker material to Carling Brewery of Cleveland. Carling was passing out wall posters, bumper stickers and some other items supporting a national safety drive. The bumper sticker read "Slow Down":

You were supposed to put this sticker on your bumper and the guy behind you would slow down....We printed ten million bumper stickers...My order was for—I think—100 million square inches....It was fifty cents a sheet—I remember that—and it was a 36 inch by 56 inch sheet. And we printed about 48 up on a four color press. We couldn't sheet the stock because we didn't have a sheeter big enough.[86]

Mahnic remembered some of the early problems at Fasson. Removable labels that refused to be removed, a large order of bumper stickers that could not be separated from the backing, truck and boat signs which would not stay put.

B.F. Goodrich had been buying Kleen-Stik for ten years when Mahnic finally made a sale of about 30,000 sheets. Soon there was a call from the man who ran the company print shop. Unlike the Kleen-Stik product, these sheets were supposed to be fanned and powdered with talc before feeding. They had not been dusted and were sticking together.

So he's screaming and cussing...and we have maybe 80 cartons of stuff in there...so I called Painesville and they would say, "hold their hands." I said, "That doesn't solve the problem." So they sent...Ted from Quality Control....So he drives to Akron with his station wagon and he goes in there and he's getting awful huff and puff and they're hollering and jumping all over him so he comes [to me] and says, "Boy, if that guy running the printing department...gives me any headaches, I'm going to punch him right in the mouth." And I said, "That's going to help a lot." He goes back and doesn't solve the problem, so Morgan tells me, "Sometimes, you know, that's our problem. We don't know enough about this stuff."[87]

Eventually Ted from Quality Control was forced to haul all the stock back in his much overloaded station wagon.

IT WAS FUN

That's how Ed Mahnic describes his years in the business.

I enjoyed it because I was in on the ground floor. I knew it was going to happen and I saw it and it was exciting. It was fun. I mean I enjoyed going to work.[88]

RAH! RAH! RAH!

Odds are that Henry Anderson is the only ex-National Football League player to run a label company. At Northwestern University, where he studied engineering, Anderson played three years with the Wildcats, and was voted All American playing guard in 1929 when he was a member of the first team All Big Ten and All Western teams. He played one year with the Chicago Bears:

I played with the Bears because I really like football. I liked the contact, the hitting and the tackling. I wanted to give it one more try, but realized it wasn't what I really wanted to do.[89]

Salem Label is actually a very old company, founded in 1862, and claimed to be one of the originators of gum labels in the U.S. But, when Anderson purchased it in 1950, it was still "a pretty small operation" producing "gum labels and just plain paper labels."[90]

Anderson says, "I think we were one of the first people to recognize the potential of the narrow web flexography process."[91] Salem had been a good customer of Mark Andrews for a number of years before this purchase. Henry Anderson had first seen a display of the tape printer at a packaging exposition in Atlantic City and ultimately owned twenty or thirty tape machines, mostly producing price and specialty labels for supermarkets and bakeries.

I remember when the first press came. When Mark Andrews, Sr., got the press in the back of his station wagon. We bought the press, we bought the die, which was made by Richard Rosemann (he was still working for Mark Andrews, I believe). And the paper we got from Fasson...we're talking '56 here....[Andrews set up the press] and then he left.[92]

Anderson's background in engineering, chemicals, plastics and sales were all useful to him as he built the business. Salem became a center for laboratory testing for Fasson in the late 1950s.

NO NICE YOUNG MEN AVAILABLE

Don McDaniel started in the business as Salem Label's first salesman in 1956. Before that he had sold insurance and farm feed. McDaniel says Henry Anderson, the president of Salem Label, was looking to hire a nice young man to help in sales. "He couldn't find one so he hired me instead."[93] Tom Wilson joined Salem Label as a printer in the same year as Don McDaniel. They sometimes rode to work together. Wilson remembered how difficult it was to get good rubber plates. "We had a little plate maker and we'd set the type...and then we made the plates from that."[94]

The rewinder was also difficult. "Sometimes the [motorized] rewind didn't want to roll stuff up and you even almost had to help it sometimes."[95]

Jim Wilson, who was also at Salem with McDaniel and Tom Wilson, says, "We had to grind our own blades from sheeter blades and make a butt cutter out of it. And sometimes it would cut through."[96]

FIFTIES FOOTNOTE

Many of the well-established tag and label manufacturers were passing on their leadership from father to son in the 1950s. This business is extremely family-oriented, and many of today's owner/operators are third, even fourth generation. A major change in a key company occurred in the late 1950s. Bill Eiseman of U.S. Tape and Label recalled meeting Mark Andrews, Sr., at a packaging meeting.

> *He said to me, "Bill, you got any kids in the business?" I said, "No, they're too small." And he said to me, "Well,...I want to give you some advice....When your son is old enough to come into the business, let him make his own mistakes. Don't tell him what to do." [I asked] "Why do you say that?" He said, "I almost lost my son in the business. He came in and was doing some things and I never would let him do anything on his own....One day he came in to me and said, 'Dad, I'm leaving...I'm going somewhere else.' [Mark, Sr., said] 'What are you going to do?...We've got this good business here, why do you want to leave?' He said, 'Because anytime I do anything you're always reprimanding me or saying it's wrong. You never back me on anything. You want to run everything, and...I don't believe I have a future here with you.'"*
>
> *[Andrews, Sr. asked for some time to think about it and went to his son and said] "Look, if I back off and let you run things the way you want to do, and make your own mistakes, will you stay with me? I really need you." And he said, "Well, we'll try it." So he stayed with me and it worked out.*[97]

Incidentally, Bill Eiseman's son, Jim, is in the business. Eiseman took Mark Andrews' advice.

CHRONOLOGY OF APPLICATIONS, 1955-1960

1958 EDP Data Transfer Labels/Picking Tickets/Routing Labels

CHRONOLOGY OF INVENTIONS/ INNOVATIONS, 1955-1960

1958 Hardened Steel Rotary Dies
Automatic Roll Label Applicators

THE MAN WHO WOULDN'T QUIT, PART THREE

Did you know Burt Morgan tried to buy into Kleen-Stik? Harold Scherer says he sat in on a meeting that took place probably (but not certainly) after Morgan's departure from Fasson:

> *But Morgan wanted to be too controlling [and Kleen-Stik wouldn't sell.] Anyway Morgan decided he'd start on his own. He did. He became quite competitive to us.*[98]

[1]Clark. The First Fifty Years, p. 98.

[2]Clark. The First Fifty Years, p. 70.

[3]*Ibid.*

[4]Burt Morgan interview with Bill Klein, p. 4. The incorporation papers were filed in New York state.

[5]*Ibid.*

[6]*Ibid.*, p. 5.

[7]*Ibid.*, p. 6.

[8]Clark. The First Fifty Years, p. 72.

[9]Morgan interview, p. 7.

[10]Clark. The First Fifty Years, p. 75.

[11]*Ibid.*

[12]Morgan interview, p. 8.

[13]*Ibid.*, p. 9. Morgan says 20 inches in the interview, but that would have created a 50-inch machine.

[14]*Ibid.*, p. 10.

[15]*Ibid.*, p. 11.

[16]Morgan interview, p. 21.

[17]Clark. The First Fifty Years, p. 83.

[18]Morgan interview, p. 21.

[19]Clark. The First Fifty Years, p. 73.

[20]*Ibid.*, p. 73.

[21]*Ibid.*

[22]Chuck Reed interview with Bill Klein, p. 7.

[23]*Ibid.* p. 9.

[24]*Ibid.* p. 14.

[25]*Ibid.*

[26]Morgan interview, p. 22.

[27]Al Norman interview with Bill Klein, p. 35.

[28]It was still Avery Paper at that time.

[29]Bill Donovan interview with Bill Klein, p. 4.

[30]*Ibid.*, p. 4, 5.

[31]*Ibid.*, p. 5.

[32]unpublished manuscript from Al Norman, September 3, 1993, p. 5.

[33]*Ibid.*, p. 4.

[34]*Ibid.*, p. 6.

[35]Buchta interview, p. 2, 3.

[36]The idea of vinyl siding was ahead of its time, and even today's vinyl siding is not fastened by adhesives.

[37]Clark. The First Fifty Years, p. 77.

[38]Morgan, Start At the Top, p. 80-82.

[39]Clark, The First Fifty Years, p. 77.

[40]Morgan interview, p. 11.

[41]*Ibid.*, p. 13.

[42]Morgan, Burton D. Start At the Top. Hudson, Ohio: Summit Publishing Co., 1982, p.74.

[43]Morgan interview, p. 14.

[44]Donovan interview, p. 8. He also thought that this was "probably Burt Morgan's greatest contribution to the industry."

[45]Clark. The First Fifty Years, p. 78.

[46]Herb Smith interview, p. 16 & 17.

[47]Bill Klein, in Bud Gray interview, p. 6.

[48]*Ibid.*, p. 7.

[49]Muir interview, p. 1.

[50]*Ibid.*, p. 2.

[51]*Ibid.*, p. 4.

[52]*Ibid.*, p. 5.

[53]Dick Riley interview with Bill Klein, p. 2.

[54]Collons interview, p. 18.

[55]*Ibid.*, p. 17. Not Rube Goldberg, by the way.

[56]*Ibid.*, p. 18.

[57]Schwartz interview, p. 11.

[58]*Ibid.*, p. 8.

[59]Ferris interview, p.2.

[60]*Ibid.*, p. 6.

[61]*Ibid.*, p. 6, 7.

[62]*Ibid.*, p. 8.

[63]Carlson interview, p. 6, 7.

[64]Dick Egan interview with Bill Klein, p. 3.

[65]*Ibid.*, p. 3.

[66]*Ibid.*, p. 4.
[67]*Ibid.*, p. 4, 5.
[68]*Ibid.*, p. 3.
[69]*Ibid.*, p. 4.
[70]Lee Paul interview with Bill Klein, p. 2, 3.
[71]*Ibid.*, p. 3.
[72]*Ibid.*, p. 4.
[73]Egan interview, p. 8.
[74]Stan Fulweiler interview with Bill Klein, p. 3.
[75]*Ibid.*, p. 4.
[76]*Ibid.*
[77]*Ibid.*, p. 6. Thanks to computer chip technology, such cards are now common. This is another example of an invention before its time.
[78]*Ibid.*, p. 7.
[79]*Ibid.*, p. 7, 8, 9.
[80]*Ibid.*, p. 8.
[81]*Ibid.*, p. 9.
[82]*Ibid.*, p. 10, 13.
[83]Fasson was a product name at first. Originally the Fasson product was Avery Paper's version of Kleen-Stik.
[84]Mahnic interview, p. 5.
[85]*Ibid.*, p. 6.
[86]*Ibid.*, p. 9.
[87]*Ibid.*, p. 17, 18.
[88]*Ibid.*, p. 20.
[89]Henry Anderson, quoted in the Salem News, Thursday, July 9, 1992.
[90]Henry Anderson interview with Bill Klein, p. 1.
[91]*Ibid.*, p. 13.
[92]*Ibid.*, p. 7.
[93]Don McDaniel interview with Bill Klein, p. 7.
[94]Tom Wilson interview with Bill Klein, p. 2.
[95]*Ibid.*
[96]Jim Wilson interview with Bill Klein, p. 5.
[97]Eiseman interview, p. 18, 19.
[98]Scherer interview, p. 10, 11.

SECTION FOUR
LIFT OFF: 1960-1970

EXPLOSIVE GROWTH

By 1960 there was tremendous momentum built up in the industry. Large material suppliers such as DuPont, Owens Corning, Dow Chemical, Shell, Kimberly Clark, Crocker Burbank, Scott and others began to re-engineer products or to design new products especially for pressure sensitive adhesive roll label production. New equipment and supply firms were incorporated to serve what had formerly been a specialized niche of the printing industry. These material suppliers were constantly expanding plants, adding production lines, adding new branch plant locations.

There seemed to be no end to the demand for pressure sensitive labels. Though the general economic conditions changed, the overall demand for pressure sensitive labels kept on growing. Between 1960 and 1970 the sale of pressure sensitive adhesive roll labels compounded at double digit rates, from about $40 million in 1960 to more than $80 million in 1965, reaching almost $240 million in 1970!

The number of competitors in the industry grew almost daily, from a dozen or so in 1952, to about 100 in 1955, 250 in 1965 and 1,000 in 1970. While some were printers who expanded into the roll label business, the bulk of the new competition came from new entrants into the industry, first self-wound pressure sensitive tape printers who converted to the laid-on label, then base material, packaging or former PSA roll label salesmen who saw the opportunity to become successful in a business of their own.

INFAMOUS THIRD GENERATION

Bill Muir joined his father's firm, Grand Rapids Label, in 1959. After his father's death in 1963, he took over the operation. He remembers the early years:

> *It was an additional label to a product—a feature label, a sale label. That's for a fact. It wasn't until a number of years later that it really became a viable alternative as a primary label....The government helped over the years with their additional requirements for information for consumers. And the easiest way to comply immediately...was to add another label.*[1]

Muir, who calls himself "the infamous third generation," recalls buying a Mark Andy press in the early 1960s, and says he could never get the Kleen-Stik product to run on the rotary press, although it worked well on their New Eras.[2]

A FRIEND IN NEED

John Beck, who founded Artistic Packaging (now API Graphics) in 1965, had a lot in common with Stan Avery. He and his wife Rachel took $500, their life savings, and started his business. In his first two or three years, his son Andy recalls, he worked probably 360 days a year and collected no salary at all. Rachel supported the family with her job as a secretary and worked nights on the books.

"There were some uneventful Christmases," Andy Beck recalls.[3] The first customer of this North Carolina venture was Burlington Industries, still a client today. By 1969 the company was doing well, but John Beck suffered a heart attack. Andy remembers:

I was in college. The only person here in the business was my mom. Jim Carr from Webtron took a week of his vacation, came down from Illinois, and worked back in our shop, running a press, that sort of thing, just to help us out until my dad got on his feet again. Didn't do it for any money, did it because of friendship between my dad and him.[4]

LIKED THE LOCATION, THE PEOPLE, THE OPPORTUNITY

Dick Pearson liked the prospects at Avery Label when he joined them as director of marketing in 1960.

I was attracted to an enterprise in southern California, so location was a factor. I was attracted to Stan Avery and Russ Smith as individuals. I thought they were the kind of people I would like to be associated with. And thirdly, I was attracted to what seemed to me the strong growth potential of the industry.[5]

Pearson, a Harvard MBA graduate, had worked at General Foods, an accounting firm, a consulting firm and Forest Lawn Cemetery before starting his career at Avery, which ultimately led to the presidency of the company. The corporation (including Fasson) was doing about twelve million in 1960 and was at two billion when he left thirty years later. "I would guess the label business was about eight million and the rest...things like Fasson. Europe wasn't very large [then]. And we had some license fees...[6]

Pearson says that "one of Avery's strengths is that they saw the potential in the business."

I think if they had been located some place other than the west coast they would have been even bigger....The bulk of the business has always been east of the Mississippi. But...from 1935 to 1960 Avery didn't have a pretense of making labels other than in California....I think...that had allowed a number of competitors to get fully established...right in the middle of the market as opposed to being two or three thousand miles away.[7]

Pearson believes they were looking for someone with consumer packaging experience because John Torrey, who was running the label operation at the time...

wanted to build a label business [in the packaging field to go along with] the nice but small stationery office products that we had. [They were the products] most people had heard of from Avery as from Dennison....They [the office products] had a lot more exposure than the custom label business calling on industrial users...[or] Fasson, which was using a different name and calling on a fairly limited group of converters.[8]

BIGGEST DRIVERS

Pearson says:

I think customers have really consistently driven the industry by demanding solutions to certain problems and offering at least part of the solutions themselves....I think they are the biggest drivers of new applications by presenting a problem [to the] converters.[9]

Avery had to have good research and development programs early on because it wasn't until the late 1970s and early 1980s that the industry grew large enough for the large chemical and material suppliers to start focusing their research and development on the needs of the pressure sensitive material producers and their roll label printer customers.

Avery has had a lot of R & D people that did important things in coming up with new materials and ways to print on them...and inks that would work....We had ink chemists who did most of the work. Ink people know a lot about ink but they really didn't know how to make ink work well on a sandwich. A big problem with PS... is you're printing on a soft sandwich and it's a lot more complicated than just printing at high speed on a very flat surface.[10]

THIRTY-YEAR MAN

On looking back at the 1960s, 1970s and 1980s with Avery, moving from marketing director to vice president general manager, then group vice president of converted products, then executive vice president and finally president, Pearson says:

As Avery grew, for me there was a lot of satisfaction in our increased ability to attract more and more talented people, the contribution they could make to making the business a bigger and better one than anyone envisioned. [In 1960] I sure as hell never thought—and I know Stan Avery and Russ [never did]—[that Avery] would ever end up at a 100 million, let alone a two or three billion dollar business.[11]

LIMITED AUTOMATION

As long as labels were applied by hand, the industry had important limitations. With automatic application, the industry began to grow explosively. In the early 1960s, Avery had three dispenser models. Don Foukal, who joined Fasson in 1961 (after being General Sales Manager for Elizabeth Arden) recalls going to California to see their automatic labeler. It was, he remembers, "a really unbelievably crude device [which utilized] a solenoid for an impression mechanism."[12]

His competition at the time was tough:

Label-Aire and Tuck...both were very good. Tuck [would] apply a label anyplace within a two inch vertical distance. The Label-Aire, because it blew the label on, didn't have any problem with an exact distance. Ours, because of the solenoid...had to be an exact distance or you would burn up the solenoid, or you wouldn't apply the label.[13]

IRONY OF IRONIES

Foukal's first sale of an Avery/Fasson labeler was to a record distributor, Dave Mages, the man who would later produce the Webtron line of flexographic presses! Foukal and Chuck Reed were selling the concept of putting the dispensers on line. It worked with labeling records, but that was a very simple operation:

> *A company in Mentor [Ohio] called Automation Development...[was] willing to design systems for us and helped me with the quotes and we got into the systems business....In truth it was the only way we could use our piece of equipment....In the final analysis it was really the best thing that ever happened to us, because adversity brought us into the system business and automation development.... At one time we owned 99% of the luncheon meat labeling business. Oh, we had it all.*[14]

Fasson began working closely with label printers.

> *If a label printer had an application the Fasson salesman would go with the label salesman to the end user. And they would complete this [automatic labeling application] form—the speeds, the labels, the assembly, the product—the whole nine yards. They would send that back to us, at which time we would log it in. For example, if an application came in from Salem Label for Superior Meats, we would log it in. If [the same] application came in from [another printer for a quote] we would say, sorry we are engaged on that application.*[15]

This encouraged the label producers to get their jobs into Fasson early. It was an exceptionally profitable business, "because capital equipment is not a price-sensitive business."[16]

In spite of the machine operation turning $450,000 gross profit on a million dollars in sales, Foukal was fired by Bill Dodge because they disagreed about the marketing philosophy. Dodge was an advocate of the razor/razor blade theory. Give the machines away and sell the labels. Foukal believed in selling the machine at a good profit, then selling the labels for even more profit.

As did many others who left Fasson, Foukal went directly to MACtac.

WALK THE TERRITORY

Jim Kyte believes Coated Products developed its first acrylic adhesive about 1960. "We had a little company, a little polymer outfit by the name of Catalin in Edison, New Jersey, that did some polymerization work and had been working with Air Products on acrylics." [17]

Kyte discovered in an interview with Catalin's research and development director that adhesives could be modified easily in the polymerization process. This produced better adhesives and specialized adhesives as well, such as those for high temperature applications.

Coated Products had "always" done foils.

> *We had one customer in New York by the name of Cameo. They were excellent old German engravers. [They did] embossing plates. We sold them a lot of foil. There were a whole slew of guys...all in the downtown area of New York City. All from 14th Street south. I used to be able to walk to most of my customers in New York City. I would be able to get on the 7th Avenue subway. I got off the ferry boat [from New Jersey] and walked around Prince Street [where] there was a company by the name of General Trademark...and J.L. May was there and Cameo was there and Foil Craft....Cameo was the leader. They were the guys who did the nicest job. We worked very closely with them on laminated foil and some plain foil....Acetate was the big thing of the time, and vinyl, and then along came Mylar™. Mylar was the big thing because it could be metalized.*[18]

One of Coated Products' more successful items was a cast vinyl film product specified by the government for aircraft marking. Only Coated Products and 3M were qualified. "So we priced our product ten percent below 3M...and it was a license to steal....Then Fasson came along and that was the end of the ball game." [19]

Kyte recalls they were coating on a 54 inch machine with speeds of twenty feet a minute, later upped to ninety. Later, a coater was added that was capable of 300 feet per minute, a quantum leap.

THE LAST HURRAH

California, probably since it is more worried about pollution than most states, has the toughest environmental laws. So, in 1967, recalled Art Schwartz, Kleen-Stik closed down its Los Angeles plant.

But just before the closing a new use was found for pressure sensitive materials:

> *[The Richmond Corporation] bought polyethylene coated kraft and things like that that nobody wanted...a room full of trim seven to eight inches wide....So I went out and we sat and talked about this thing. [Arthur Creatura said], "We make polyethylene bags, we have... the printers, we have flexo printing. If we could put pressure sensitive on this stuff...laminate another piece of kraft...we could cut it off and register and make an envelope that would have a window." We worked out the specifications between us...we ran a few samples and developed a specification and all of a sudden we got orders for 100,000 feet...another 100,000....And just before Kleen-Stik went out of business, I turned in what I was informed was the largest single order for pressure sensitive that the corporation had ever received from one particular customer. It was for three car loads of 20,000 pounds [each].*[20]

MARRIAGE AND DIVORCE

By 1960 the pressure sensitive adhesive business finally attracted the attention of corporate America. Jim Dillon, a senior vice president of National Starch, persuaded his company that Kleen-Stik would be an excellent acquisition. Dillon, ironically, became aware of Kleen-Stik because of a lawsuit filed by the company against National Starch claiming its 26-2404 adhesive had failed. The lawsuit was lost, but Dillon became fascinated by Kleen-Stik's ability to purchase large volumes of acrylic polymers. Al Norman says the purchase of Kleen-Stik was concluded in 1961.[21] Kleen Stik by then had plants in Chicago, Newark, Los Angeles, Toronto and Mexico City.

National Starch and Kleen-Stik should have been a marriage made in heaven. National had money and much technical knowledge, especially in solvent coatings. Kleen-Stik was still having release problems in 1960 when they were given the lion's share of John F. Kennedy's bumper sticker business.

It started off well. George Collons said, "It was a hands-off deal. They gave us money and said, 'We're not familiar with what you've got here. You go ahead and buy where you have to buy.'" [22]

Unfortunately, Jim Dillon died of a brain tumor within a few months of the friendly takeover. Without Dillon, National's interest in the company waned. It became an orphan. The "two Jerry's" (see sidebar "G or J") were eased aside and Gerry Cole was killed in an automobile accident in 1965.

In late 1968 Garrison Brinton and John Dickenson acquired Kleen-Stik and the Ludlow Corporation's Reinforced Tape Division and created Compac Corporation. "After a couple of years of successful operation they floated a stock issue and they took the money and built a new factory." [23] The old plant was to be shut down on a Friday and the new one opened on Monday. The shut down was effective, but the new plant had terrible problems. Art Schwartz remembers a twenty percent scrap rate![24] The coating portion of Compac was later sold to Essex Chemical, which dissolved the company in the early 1970s because it was unable to meet the environmental standards required to run solvent-based adhesives. Label-Aire division was retained and set up as the Label-Aire Company. (National Starch was ultimately acquired by Unilever.)

G OR J?

Kleen-Stik was essentially the product of two Jerry's—Gerry Cole and Jerry Zalkind. It had started in 1946. Wayne Lasinski remembers:

> *Our deal was versus Avery all the time and even when they started up Fasson...it's a fact that we did not compete with the roll label printers. And didn't produce any printed labels.*[25]

Gerry Cole was president and super salesman, and Jerry Zalkind was another super salesman. Zalkind "contracted polio the year before Jonas Salk developed a vaccine. So until he passed away he was in a wheel chair....He was okay from the waist up and when he came into a room for a meeting, after 30 minutes you didn't see the chair." Art Schwartz remembers Zalkind as "absolutely fantastic." [26]

> *He was something else. He was the fellow who looked at the Monarch [Marking System] ticket and said, "They ought to have a pressure sensitive ticket," and he personally sold them the Kleen-Stik/Monarch relationship which was responsible for a significant portion of Kleen-Stik volume over the years....Jerry targeted Monarch as the largest pricing company in the business and they were.* [27]

Monarch became one of Kleen-Stik's largest accounts.

BASE MATERIALS

The base material suppliers were doing their part to come up with new and better quality pressure sensitive adhesive coated materials to meet the growing needs of the roll label printer. In addition to improvements made in release liner technology, the paper and film manufacturers were beginning to look at the rapidly growing pressure sensitive adhesive label market as a growth market for their products. They began to improve the quality and consistency of the materials sold to the adhesive coaters. A greater variety of adhesive/face stock combinations became available to meet a greater and greater variety of specific label application requirements.

Not only were there now more adhesive coated products available, they were increasingly more consistent and of better quality than the earlier materials. Because of the greater scale of operations and aggressive competition among suppliers, the price for the improved materials was equal to or lower than prices a few years earlier.

PLEASE RELEASE ME

In 1963 when John Questel picked up his sheepskin at the University of Akron, a Master of Science in polymer chemistry, it was somewhat ironic. Two years earlier he had helped get the sheepskin out of the PSA business.

The original release liners for pressure sensitive papers were parchment or glassine. Parchment paper is cotton treated with sulfuric acid, which makes it very dense and almost impervious to grease. Glassine is made from wood stock and is still common in window envelopes. Both are brittle and tear easily. Questel helped switch MACtac to a super calendered Kraft paper when he was the sole bench chemist at Morgan in 1961, he recalls, and the entire industry followed suit.[28]

> *Good decision. Much higher tear strength. Parchment papers...would tear so easily. The moisture in it was unstable [so] the printers would get more splices than they could handle...as the paper absorbed moisture.*[29]

> *The net result was a much more trouble free laminate on the printing presses and in the adhesive coating machine.*[30]

When Questel left Morgan in 1966, he formed Adhesive Consultants, Inc., which "has assisted 24 different companies throughout the world during their foundation period [enter the pressure sensitive adhesive coating business]." [31]

He has helped many companies install environmentally friendly ways to produce base stock in the United States, Columbia, Greece and South Africa. In the 1980s his firm led the way in "the production of monolithic transdermal drug delivery systems for nitroglycerine and other medicants."[32]

EXCEPTION TO THE RULE

So far in this history there have been very few women mentioned. The use of "salesman" has been deliberate and accurate, not chauvinistic, because the first 25 to 30 years of the industry was completely male-dominated. Also, most comments made by one member of the industry about another have been uniformly polite. Even Stan Avery and Burt Morgan were charmed by the other.

Both chauvinism and politeness are about to change. Meet Bev Eagon from Dow Chemical who joined Fasson in 1962.

YOU'RE NOT HUMBLE

Stan Avery was still innovative even after leaving Fasson in the hands of Bill Zimmerman. But either while at Fasson in 1959 or a few years later he had given the company a mission—to develop a pressure sensitive adhesive product with a split or cut in the sheet stock backing paper to make it easy for the customer to remove the backing. It was to be like cracking and peeling a hard-boiled egg.

Fasson chemists were working on the project. The original concept—which ultimately proved successful—was to apply an acid in a diagonal line on the backing sheet. The acid would etch this tiny line so the customer could bend the sheet (CRACK) and the sheet would be easy to remove (PEEL).

When Eagon arrived in 1962, she looked at what was going on in product development and promptly said to her boss, Les Dickard, "There's really not much to this problem."

> *I mean technically they were just using the wrong acid. They were using sulfuric acid. Now you just think about sulfuric acid at the boiling point and you know you're sure as hell never going to get enough on the paper. He [my boss] thought I wasn't humble.*[33]

Eagon says it was quite a shock coming to this small entrepreneurial company from Dow where "you were only arguing data."

> *Everybody was jockeying for a position....I would have been smarter to say, "God, this is tough." Well, I tried a number of acids... then I went with hydrochloric and we stuck with that. Now we ran into a lot of problems but most of those were in manufacturing. But the basic problem...getting rid of the sulfuric acid...was solved in a couple of months.*[34]

One of the manufacturing problems was that the acid would sometimes work too well, causing web breaks in the ovens. Fasson's Cal Miller explained:

> *Bill Stanziale was running the coating machine. He was the one who came up with the idea to do the acid etching to weaken the backing paper after the material had been coated, rather than before. After he had crawled through the ovens five or six times every night fixing the paper breaks...he decided that there should be a better way.*[35]

Eagon says that she and the other Fasson people thought it was a "little gimmick and we were going to get customers because of the novelty," [36]not understanding the significance of the product, "but Stan Avery did." [37]

Crack-n-Peel was introduced in 1967 at a 25 percent up charge. It was an immediate success, helping to catapult Fasson to a position of leadership in the pressure sensitive adhesive coating industry.

SERENDIPITY TIME

At Fasson, serendipity was a favorite word. When things went right it was serendipitous, when they went wrong, there was no serendipity.

Eagon recalls that perhaps as early as 1964 she wanted to be coating water-based adhesives. Her boss did not agree but Bill Zimmerman, his boss, did.

> *At that time you had so much serendipity time, it meant like every week you could spend so many hours on your own projects. Anyway he was going to let me coat this water base. So I go over and I think it was the number two machine and I started working... and the damn stuff buckled. By the time it got to the first roller he [my boss] was so mad I thought he was going to fire me. And so I had to drop it.*[38]

Eagon remembers her fellow pioneering PSA chemists, Bill Chang (who worked on the silicone release liner), Glen Brewster (who helped develop the solvent-based acrylic adhesive pressure sensitive product) and S. Michael John (later at General Electric).

SOABAR, SO GOOD

MARK WERT: When I came out of college I did not know what I wanted to do and my next door neighbor was a production manager for Soabar. So he said, "Why don't you come into Soabar and take some tests to see if you want to go into the tag and label business?" ...This was 1963. And I took a test with Soabar and...I started out as a service man repairing labeling machines for Soabar.[39]

Wert said that Soabar's main thrust was in the apparel business "I always kiddingly called the rag industry...putting basic one and two color PSA labels on [products]." [40] Label-Aire was the applicator of choice until Soabar began to build its own labelers.

Then we started getting into the hosiery business, making the pressure sensitive [labels] to put on socks. How do you get an adhesive that will go onto socks that won't stain, won't leave residue. Boy, that was a whole education.[41]

In 1968 Avery bought Soabar, which had been a family-owned business. Avery kept Soabar as a separate entity, since it made its considerable profits selling complete systems. Wert said that this was unique, "We could not only talk to purchasing and marketing people, but we could also talk to engineers about...suggestions or recommendations as to how to put it on."[42]

From 1972 to 1974 Avery began consolidation with the Soabar group, reducing Soabar district offices from seventeen to ten and increasing pressure sensitive while decreasing tag sales.

I'm just old enough to remember when we used to have four seasons. Like when I worked with Soabar...there were four purchasing seasons. Things used to be purchased in quarters. Things are no longer done that way...People come up with an idea, they rush to market with it, and you have to be more promotionally geared. People don't buy to inventory anymore.[43]

Wert said he learned a lot at Soabar that has helped him in his other jobs (Tolas, MPI).

One of the best things I got, in the early 60's when I was with Soabar, was to learn quickly that everything, everyone is [either] an application or a prospect, because all products in one way or another have to be identified...so you do it with a label or you do it some other way....The biggest thing that helped me is your pressure sensitive hasn't really gone [up in price in]...comparison with the gum and glue market... but whatever the problem is you've got more labor on the other side...your applied costs. I think Fasson was the first one that came out with that formula for applied costs.[44]

In 1980, after working at Soabar for seventeen years, Wert decided he was tired of traveling and wanted to stay in Philadelphia. He moved into pharmaceuticals, joining Tolas. Tolas had begun as Fenton Products in 1898. Later Fenton's son-in-law changed the name to Tomkins Label Company. Tomkins Label was one of the original Avery distributors in the 1930s. By the time Wert joined the firm, it was called Total Labeling Systems (Tolas), specializing in the pharmaceutical market. They were a forerunner in using Tyvek, said Wert:

They would get it from DuPont. DuPont would send it to [a PSA coater] who would coat it...and then Tolas would sell to a lot of pharmaceutical folks....They were in a little different market.[45]

You've got to sell value added....When I was with Soabar I used to put on seminars on what price really means....What is it that you're trying to accomplish?...What are they really looking for?...A price is always secondary, if you can help someone sell their product.[46]

THE SOCK FIASCO

This story runs around the industry. In fact, there are so many stories it is likely the problem occurred more than once.

Socks are labeled and moved to shipping stations on hangers which essentially "pair" two socks. Gum labels worked well to hold them together, and pressure sensitive labels were even easier to apply. But they didn't always come off. The joke was that Simon Adhesives salesmen (or some other company—the stories are not always clear) wore socks with labels still on them, because the company had to eat the mistakes. MACtac is also implicated. One of the companies that admits being involved in a sock label fiasco was J.C. Penney, which claimed the labels supplied by Adhere Products were to blame. Dick Egan stated that:

> *We sent Pat Mehandle to New York to live and call on the [sock] brokers in the Empire State Building. The 21st floor was made up of men's sock brokers...and I think the next floor was all women's hosiery. I got a phone call from Pat, and he said, "Are you sitting down?...I'm in Brooklyn, in a warehouse, and I'm going through the J.C. Penney's inventory, where they had taken the bands we had furnished them and applied them to socks, and I can't get them off. That adhesive is just frozen to the sock."*
>
> *It ended up $120,000 damage [claim]....The base price on the sock was probably fifty cents, so you can divide that into how many pairs of socks were lost. We were buying the material from MACtac Adhesive. It was supposed to be a special type of adhesive.... Mostly the socks were made out of nylon and of course there was a reaction with the wrong adhesive. [So] they picked up part of it and we...picked up like $20,000....That was probably the worst scenario that I had got in.*[47]

Egan says it took almost three years to get the adjustment

Now "we do a [label] that goes on a pair of socks and it's got Cheryl Tiegs' face on there....Twenty years ago we wouldn't have even attempted to do something like that," [48] says Andy Beck of API Graphics.

DO-IT-YOURSELF

Williamson Adhesives of Skokie, Illinois, came up with an innovative idea in the early 1960s (approximately). Sticky-back cellophane tape was quite expensive then and represented a major office expense. So Al Sobel of Coronet Paper began to sell Williamson's sticky tape-maker.

The machine took plain cellophane which was threaded through the machine under a coating head which was attached to a quart can of adhesive. Pull the tape through and it coated it.

If you pulled an inch, it coated an inch. If you pulled a foot, it coated a foot. And it had a little air-dried festooning arrangement in there. It worked beautifully and I really had fun with that one. I took the machine to Permacel and they took me to all the Minnesota's [3M] accounts.

Then I turned around and showed it to Minnesota and they took me to all of Permacel's accounts...and they bought like mad. With the advent of that, the price of [sticky] cellophane dropped to a reasonable level and then [the desk-top coated] was no longer needed. But it did cut the cost of cellophane tape in half.[49]

This experience, however, led Coronet into the self-adhesive decorative bow business—the stick-on bows for Christmas and other gift packages. The early 1960's machine had one head and a magazine of pressure sensitive cards. It was very slow and it required an operator as well.

So I said, "Well, why don't we work from rolls?" ...I would print it and slit it. And my source of supply became unreliable, so then I had to get into the adhesive coating business to fulfill the needs of the bow industry.[50]

Now much faster two-headed machines are employed and one operator can service ten machines.

THE USER FRIENDLY PRESS

The introduction of the first "user friendly" narrow web flexographic press was a major PSARL achievement. Oakland Products was a design and manufacturing shop owned by engineers Bob Holyoake and Frank Moynes. Says Duane Polkinghorne (later to go into business on his own), who joined the firm as purchasing agent in 1967, "You came to this firm with an idea, they would design it...and manufacture it."[51]

In the early 1960s, perhaps 1963, they were working with David Mages, developing a four inch wide Webtron press, called the "Dart." A later four inch was the "Flexo Master." [52] Polkinghorne says the original four inch machines were not especially popular although the design was revolutionary at the time. "They sold maybe one a month." [53]

Leon Beaudoin, who had learned how to run Mark Andy's for Doring Label in the late 1950s, joined Oakland Products in the early 1960s. Mages wanted a man in the east and gave Beaudoin a ticket to come to Chicago and look at "the factory."

Unbeknown to me the plant did not belong to Webtron at the time. He was subcontracting. He didn't tell me that wasn't a part of the company... [which] only consisted of a two room office on Peterson Avenue.[54]

When Beaudoin joined Oakland Products the only Webtron was the $4^1/_2$ inch press. But the 650 was ready to be introduced at a trade show in 1967. The 650 ($6^1/_2$ inches wide) was, as was the smaller Webtron, built with mostly off-the-shelf equipment incorporated into the design.

We built two machines. One ran, one wouldn't. They were built side by side, identical parts...and everything else....One performed and one did not. We intended to have two machines at the show. Dave Mages, being a marketing type guy, was really sharp. He decided to take a van that had been displaying the four-inch machine, put [the 650 that worked] inside the show and put the machine that didn't run inside the van for preferred customers. The customers would come in, have a drink, sit and chat...and look at the machine. [Meanwhile, at the show itself] out on the floor, the machine that did run was just humming along.[55]

They returned from the show with 13 or 15 written orders. Which, ironically, destroyed Oakland Products. The large number of press orders tied up the plant completely so they had to change. "They had so much business they couldn't afford to do business any more." [56] So they merged with Mages and created Webtron as a three-way partnership.

YOU OWE ME

Burt Morgan said Dave Mages was into MACtac for $200,000.

And he was real honest. "I'll tell you what I've been doing. I've been taking this money to design this press and if you foreclose on me you're not going to get anything, but if you'll stay with me, we'll pay it all back" ....And I gave him the credit against everybody's judgment. Boy, he never forgot he owed me one, which is a rare thing. Whenever he sold a Webtron press, boy, we were in there with the paper. I can remember one order out in California and this guy had all Fasson paper. [Mages] sold him a Webtron and they couldn't get it to run. So he went out there and all he saw was Fasson paper. He said, "Hey, don't you have any MACtac here?...I can't get it to run until you get some MACtac." He sent the guy over to the warehouse and got a couple of rolls of our paper and put it on and Dave re-threaded the machine, turned the rheostat to full speed, turned his back to the machine and lit a cigarette. I asked Dave, "How the hell did you do that?" "Well, these guys were running it real slow....If you run it too slow it'll break the matrix....I put your paper on it, threaded it up, started it full speed so the matrix wouldn't break." [57]

Soon the business grew dramatically, but the financial resources were not available. They went to a public accounting firm planning to go public and sell stock. The accounting firm had a customer who was looking for acquisitions. "That was GBC [General Binding Corporation], and the accounting firm said,"If you're looking for somebody to buy, buy these guys. And they did." [58]

Mages stayed on as president, Holyoake went to GBC's corporate headquarters as chief engineer, and Moynes left the business, investing in real estate. In approximately 1970 Holyoake became president of Webtron and Mages retired.

Mages, one of the true characters of the industry, was impossible to locate for this book. He ran an airport and built sailboats and some reports now have him sailing around the South Pacific.

And then, after David left we did some 16 inch work for American Can. That was under contract and we got into the wider web and of course things just began to grow. Unfortunately it was a period of time when GBC...went through four or five presidents. Things were just absolutely in a turmoil.... As far as I was concerned it never settled down after that until GBC sold it to Didde [Glazer]. That's when I left.[59]

Beaudoin says the new Didde Glazer president had a sales manager who decided to do away with all the regional managers. This happened in approximately 1987.

And they did away with Bob Wornberg, Dick Shackley down in Atlanta, myself up in Stanford, Connecticut. Of course Jim Carr had passed away in Chicago so they didn't have to deal with him.[60]

YOU'RE UNEMPLOYABLE

After the sale to Didde Glazer, Holyoake returned to Webtron, which led to Duane Polkinghorne's leaving the company. Although personal friends, Holyoake told Polkinghorne he should prepare to leave because he was unemployable. So, with about thirteen weeks of severance pay, Polkinghorne started Flexo Accessories, to sell specialty accessories for the Webtron presses:

Webtron was good enough to allow me to manufacture, without any competition, for about six years. It was a four-way adjustable printing head for Webtron. That device bought me a machine shop. We now manufacture the Propheteer Line of flexo, UV flexo, letter press and rotary screen equipment in that same shop.[61]

Polkinghorne says his roll label converter customer base is responsible for innovative improvements throughout the flexo industry:

> *Our customers [keep] demanding more and more capabilities on the press and they go out and sell the product and come to the manufacturer...and say, "I have to run this, show me how to do it." ...You know the American way has always been to create a product and then go out and say this is what fits your needs, whether it does or not. Now we've got converters out there that are smart and in many cases smarter than the manufacturers and they see innovations and they'll come to the manufacturer and say, "If you'll do this, I have a 90 percent certainty that [these are] the results that we'll get out of it." This is tremendous.*[62]

Mages would accept ideas for his new press from any and everyone. Chuck Reed of Fasson recalls:

> *Dave was the kind of guy who liked to get you into brainstorming. That sucker would keep you till midnight just talking over ideas. The next thing you knew, he was putting those ideas to work either converting or in modifications for the press....I can remember sitting with Dave when we were drawing up the plans for that thing and throwing in all the odds and ends that we thought Mark Andy should have done.*[63]

The Webtron press caught on quickly, challenging the relatively flimsy Mark Andy press. Mark Andrews, Jr., recalls the effect of competition.

> *In the sixties we began building in-line presses and we were challenged by our first big hunk of competition...by Webtron. Webtron introduced an in-line flexo press. Both companies down through the years have been tremendous competitors...and have helped both businesses...grow. We were building central impression presses historically.... They thought maybe the in-line approach was good.*[64]

Alan Campbell credits Mark Andrews, Jr., with making the Mark Andy competitive. The company was "really struggling and losing share and Webtron was just showing up every place and Mark, Jr., came on the scene and sat down and used his engineering talent to put together a line of products that allowed them to compete effectively with Webtron." [65]

Lee Carlson credits the Webtron 650 with forcing the rotary die cutting industry to go from soft dies to hardened dies. Hardened dies had been around since the late 1950s, but it was the 650 that changed the industry from labels only to converting.

A label company could use their Webtron 650 to produce a gasket, for instance. First it was basically all paper, then it started with films and films over laminates, and it has progressed to now triple and actual quadruple laminates.[66]

Carlson also recalls that teflon die coating, which came out in the early to mid 1960s, was popular in the late 1960s and early 1970s. "It's very rarely used anymore [although] it has its advantages in some cases."[67]

Mark Andy produced both in-line and central impression presses in the 1960s, and began to specialize in the in-line presses in the 1970s. As a result of the Webtron/Mark Andy competition, the roll label industry began to receive much better flexographic printing equipment. The presses were wider web, easier to set up, ran faster and could hold desired tolerances.

DREAM MIND

Ed Vandenberg was trained as an engineer in the Netherlands before moving to the United States where he joined the engineering department of Proctor and Gamble in Cincinnati. Jim Brown of National Label asked Vandenberg to look at a press he was thinking of buying. This was one of the early Mark Andys.

Brown bought the press and asked Vandenberg to test it. "One thing led to another and he asked me if I could maybe run it for a few evenings."[68]

Soon a few evenings grew to many and when National Label got really busy, Brown asked Vandenberg to oversee a two-Mark Andy operation. Vandenberg agreed. After working for a while he decided the press could be improved. "I'm not putting down those early Mark Andys...but it was indeed in the early stages of the industry and naturally there was a lot of room for growth."[69]

Soon he designed a label press of his own, a central impression two color unit which he sold to National Label. He built another for National and a third for another printer. "From there on I went into business in 1962 building label presses [as Roto Press Engineering]." [70]

The original Roto Press Engineering presses were central impression cylinder type presses and printed two, three and later, four colors. The entire press was designed from scratch including the die cutting mechanism. Vandenberg introduced his press about the same time as Webtron,

The Webtron was an in-line press. Mine was a central impression cylinder. Our presses were very popular for reasons that I had intended, I invented in my dream mind. I was dreaming about printing presses and I dreamed I should do it such and such a way. And sure enough, I woke up and what it was is the methods of pivoting the plate mechanism around the impression cylinder. That made for a much easier and more scientific way of adjusting the platen to the paper and the anilox roll.[71]

In the late 1960s Vandenberg became one of the first presses to allow four color process printing, opening up the primary label market.

GETTING IN ON THE GROUND (DIRT) FLOOR

When Jim English got out of the Marine Corps at the end of 1962, the job market was bleak. But Fitchburg Coated Products offered him a job as a coating machine operator at their Moosic, Pennsylvania location. English recalls only five employees were there at the time and the floor was simply dirt. He would later work in the mixing room making adhesives, and then became a supervisor.

Being in on the ground floor I had a lot of opportunities. I had a lot of guys who were really pushing me....We had a general manager, Mike Ramano, who always tried to convince me to move ahead, make moves....He got me into tech service and new product development. So that's how I started working with customers like Alan Hollaender at that time. George [Froats of Fasson] was my competition. He took Kimball away from me. I changed the product somewhat and made it run faster and [took it away] from him. He'd change and take it back from me again. So we kind of competed like that for many years.[72]

In 1965 Fitchburg bought Simon Adhesives Products of Long Island City.

We moved that operation up to Moosic, Pennsylvania and incorporated that into our coating line. I can say the first thing I had to do was go to Webtron and go to school with Leon Beaudoin and Bob Ornberg....Leon taught me how to run Webtrons and so did Bob. So then when I sold the product, I had to go down to New York City and make it run....When we sold a product, we probably had to sell it twice.[73]

English remembers a particularly stormy meeting at Avon in Suffern, New York. A competitor, Max Freeman, was "just eating our lunch." [74] So English of Fitchburg and Chuck Reed of Fasson went to the vendor program at Avon in Suffern, N.Y., and tried to address the problems of die cut-throughs. The Avon labels were being cut on a flatbed hundreds up.

Chuck and I [decided] that when we walk through this vendor program we would just say they ought to specify [that the labels] must be cut on a 650 Webtron....We introduced the die stain test to the customer...We said you need to get a die stain on the roll and if there are cuts in it, don't run it. Then Freeman stood up and fired his briefcase at Chuck and me. But you would sell a product, twice, three times, you'd have to go out and spend all night on the presses, showing them that, yes, it was a tight release, [but] we could still strip it. Put it in freezers or use heat lamps, dry ice, make it run. [75]

THE WIZARD OF OOZE

That was the pet name for John Boynton, who helped develop adhesive formulas. English also credits Fitchburg general manager Mike Ramano and Dean Thorp, technical director, as being some of the people who helped him the most.

We did a lot of work on the coaters and different release values, silicones; we got products to run better, stronger paper to strip faster. I finally convinced Howard Paper Mills down there, Bob Finley, to make a sheet for me that was almost as strong as polyester. He did it for me and then I got the business back from Fasson. Then, after my exclusive ran out about a year later, Fasson bought from Howard and they got back into it heavy. [76]

Howard Paper Mills of Dayton, Ohio, is a strange bedfellow for any large PSA supplier. It is a small mill that could not produce a commodity paper if it tried. It is a value-added mill, producing colored bonds, offsets and many finishes. A call to Bob Finley revealed the paper involved was essentially Maxopaque, an opacified, uncoated offset. It was relatively expensive because of what must be used to create a true (not overly blue) white. Die makers were given fits with opacified papers, and sometimes complained the paper was covered with rocks. No, it is the titanium. The dies had to cut metal.[77]

English remembers Seth Wheeler who came from Fasson to Fitchburg, and helped Fitchburg get into the industrial tape market. Jim English would leave Litton Industries in 1975, partly because they gave him an assignment he felt—"even after coming out of the Marine Corps"—was just too tough. He had to fire Seth Wheeler.[78]

FROM JUKE BOX TO EDP

Many of the entrepreneurs who entered the PSARL business did so after returning from the military in their early to mid-twenties. Some did so at the age of 35 to 40, but Don Eyster started his first business when he was a junior in high school!

It was the juke box business. He spent about 80 hours a week at it, and he disliked going "to crummy bars with all the drunks."[79] So he went to Florida where he worked as a fishing guide on a charter boat. In about 1956 he began to work for Moore Business Forms as a salesman. "I spent about five years with them on computer forms and once in a while, a few labels."[80]

After five years of smaller and smaller territories, he quit and went with Woodburn Printing in Terre Haute, Indiana. After a very short time Woodburn asked Eyster if he would move to Kalamazoo, Michigan, to manage their Lambooy Label Company.

> *I told them I didn't know anything about labels except you licked them and stuck them on a box and they said, "Well, we know you're honest and we'd like you to go" ....I was up there five years as general manager of that and we made mostly embossed foil...and pressure sensitive mylar™ vinyl laminates. Pressure sensitive labels primarily for the water heater industry and things like that.*[81]

Eyster says they purchased a Kluge press in approximately 1966, "and we had the first one that had marginal punch, fan fold for all that and made data processing labels."[82]

Because of his business forms background, Eyster was very much interested in computers and realized the importance of the EDP label marketplace. Soon Eyster went with National Printing Converters in California to be close to the heart of the emerging computer industry. His first job was "to create enough business to open a plant, and it took me ten months, strictly data processing only—pressure sensitive label."[83]

After eight years, Eyster became a part owner of Data Label, Inc., which grew from nothing to almost $25,000,000 when he retired in the early 1990s.

I guess I made labels for a lot of reasons and usages over a period of time. I've made labels to label every prisoner in the state of Tennessee. I've made labels for water heaters, boats, house trailers.... I've made labels for about anything you can think of.[84]

Eyster believes that the late Eric Kline (who was with Eyster at National Printing) may have been the pioneer EDP printer:

I always hear, and I can't verify it, that he ran the first pressure sensitive data processing labels that were ever made...ran them for Alan Hollaender. He was a pressman and a very nice guy. He was always trying for something different.[85]

This is another example of concurrent development, as Avery also claims the first EDP address labels.

NICHE PLAYERS

Jerry Nerad's father founded Professional Tape Company in 1953 and incorporated it in 1955. Originally it converted pressure sensitive tape. As did most of the tape printing companies, it began to produce pressure sensitive labels.

My father created a demand for self adhesive applications primarily in the hospital environment and today we are exclusively a niche player. We are the largest supplier of hospital labelings in the country. [86]

MPI's Bud Gray says Nerad was the industry's very first niche marketer. "You can't make labels for everybody.... Jerry Nerad did that years and years ago."[87]

Nerad's company, renamed Timemed Labeling Systems in 1988, targeted the medical field because John Nerad had been a pharmaceutical salesman. One day, at Jerry's Little League game, another father present was a Doctor of Bacteriology working for one of the meat companies in Chicago. He told John if someone could come up with a self-adhesive tape to replace a gummed label, it would cut down on the risk of infection that resulted from the hands coming in contact with the mouth while licking the label. There was indeed a solution to the problem.

JUST IN TIME

As luck would have it, John Nerad was handling a "little product line"[88] which provided pharmacies with pressure sensitive tape imprinted with the name and address of the pharmacy. So Nerad returned to the company supplying the product and asked for some plain rolls of tape. He gave these to the laboratory for evaluation.

No go.

The tape could not stand up to the rigors of the lab. "It needed to be auto-claved and incubated...and subjected to water and oil baths."[89] But John persisted, seeking out help from Mystic, an adhesive coating company then owned by the Borden Company. They were able to create such a tape. Then John put his family to work stuffing samples and order blanks into envelopes. In 1954 he did $100,000 in sales.[90]

There was a demand for a printed tape that could meet the same conditions, so John Nerad had an engineer design a press similar to the Mark Andy and started printing on tape.

About 1960 the pharmaceutical companies began using plastic containers to end breakage, and the regular dextrin (water-moistened glue) labels wouldn't stick.

> *We came up with a dispenser [for PSA labels] that could fit to the top of the typewriter and that's what launched our career in the self-adhesive business. Then we designed other formats for the nursing department and the central supply.*[91]

Nerad said the pharmacy labels were printed just as the earlier gummed labels with logos and whatever else the hospital wanted. Jerry Nerad joined the company after graduating from the University of Illinois in 1964, when sales were slightly over a million dollars a year. Was it easy for him to enter the family business? "Well, let's just say my apprenticeship was not very easy." [92] Jerry Nerad says if he were to do it all over again:

I probably would have started working for someone else and come back into the business a little bit later, when I had earned the respect of my peers instead of being the boss's son, and the tough expectations of the father might have mellowed out a bit.[93]

In 1969 all of the management of the company, except Jerry Nerad, resigned *en masse* and opened up Shamrock Scientific in direct competition.

It was at that time that we brought in some new talent. [We had been] parochial in our thinking, we were comfortable because we owned the marketplace, and, by bringing in some outsiders, it changed how we did business...and created lots of new opportunities....It was probably beneficial to us because it brought a greater world around us and it really made us think differently about the business and where we really wanted to go with it.[94]

And his competition, as far as he knows, is doing okay. "There's plenty out there, so if you execute well, everybody should prosper."[95]

STILL A PROBLEM/SOLUTION

With the very real threat of AIDS confronting hospital workers, all the lab technicians, especially those involved in blood work, wear latex gloves. The labels in use at the time would stick entirely too well:

The latex would rip. Gloves were in short supply. As a result we kind of pioneered what we call a "glove adhesive." It's still a permanent [adhesive] but with a very low tack initial set up, so the technician could remove the backing with the fingertips and be able to apply it [safely].[96]

Nerad says glove adhesive is now an industry-wide standard for laboratory labels. He also added, "Fifty percent of our business is coming from new applications that we've launched in the last half a dozen years."[97]

Timemed Labeling Systems supplies hardware, software and labels. "I would say that probably two-thirds of my business is computer-generated label formats exclusively for health care."[98]

STICK IT TO ME

When Bud Gray graduated from Indiana University in 1965 he went right to work selling for Uarco Business Forms and quickly decided pressure sensitive labels were a good deal.

> *I sold a lot of business forms but what was unique about the labeling side of the business was that rather than paying a five percent commission...they paid fifteen percent for selling labels. And there was nobody [else] selling labels. So this was pretty neat stuff.*[99]

Uarco's labels at the time were fairly simple, mostly blank, or one or two color EDP labels. Printing was done on New Era flat bed letterpresses. They were slow and rather limited, mostly uncoated litho paper with a permanent adhesive.

> *We found ourselves...explaining to people how we got the materials sticky....They would look at it and stick it to their hands, they would stick it to their chest, they would stick to anyplace they could.*[100]

Gray remembers that Kleen-Stik, Uarco's chief (perhaps sole) supplier, was not particularly consistent. Some he says, couldn't be pulled off the liner sheet with pliers.

> *On my very first job they were imprinted on an IBM 402 accounting machine...[which] used real goofy spacing—5/32 of an inch spacing. There was a blank spot in the center of the label that wouldn't print. Most all the labels we sold were...about three inches wide and about one inch tall....Most of my applications were for identification purposes. My very first were... placed on shoe boxes full of IBM tab cards.*[101]

WHERE'S YOUR LIFT TRUCK?

In 1968 Dan Tomlinson, now of Labelcraft Products, Ltd., in Ontario, Canada, was living on his wife's salary, driving his wife's car, but somehow he came up with $5,000 to buy a one color Webtron. The fact that he didn't know how to run it didn't seem an insurmountable problem, but when the press arrived, there was a big problem. The driver asked where Tomlinson's lift truck was.

I had to run down the street to a lumber yard and they were good enough to come up and take it off [the truck] for me. I couldn't get it through the door so I had to sleep in the car that night [to keep it safe. We] moved it in, put it on the floor and I was so proud of it and polished it up and bought blue ink and had some plates waiting for it...and a one by three die....My wife came over and I said, "Isn't this beautiful?" She said, "How you gonna run it?" And I said, "No problem." Put paper on, threaded it through...filled the ink up and turned it up slowly and it's running along at fifty feet a minute. She said, "That's wonderful." I said, "You think that's great, watch how fast it goes." Right up to 300 feet a minute, the ink went flying in every direction, all over the plant, my wife, [everything.] That machine is still stained.[102]

NO MONEY, NO BUSINESS, NO CUSTOMERS—GO FOR IT

Tomlinson's early goal was to be a fine artist. Painting was satisfying, but it didn't produce much in the way of income. He answered a want ad and got into printing polyethylene. He went to college at night and was one of the two top graduates, which, in 1964, led to a job with a large packaging company as assistant product manager.

Sales increased and he was promoted rapidly, a salesman briefly, then product manager, then brand manager, sales manager and finally, manager in 1967. Sales rose from $200,000 to over a million, but Tomlinson was unsatisfied:

It was an English-based firm....They had a philosophy of dividing the company in two—the manufacturing arm was one part of it and everything else was the other part.[103]

Although Tomlinson could sell the product, he would not always get it delivered from the other half of his company and he was required to buy from them. He says everything became very political and people were cited for leaving early, "seven or eight o'clock every night."[104]

A deal to head up a Canadian operation for a U.S. firm fell through. He says he would have jumped out of the window, but he was only on the third story. Discussing the problems (no money, no business, no customers), his wife told him to go for it. His first customer was a major bread company in October of 1967 (not peel 'n seal). Christmas came that year when a check for $97.50 arrived on December 23.

Tomlinson's second big break occurred when his former employer "really screwed up a big job for Colgate Palmolive and [their] purchasing agent sort of searched me out."[105] Tomlinson solved the problem and the purchasing agent said he could bid on Colgate's business. That's when he bought the Webtron.

> *And from there on I would go to Colgate in the morning, [get] the "five cents off" stickers. I'd rush back and order the plates, and then I'd run off and deliver the next morning. So I got into that philosophy of service and being able to move fast.*[106]

An emergency call from a cigarette manufacturer led Tomlinson to his first mega-order. He had to contact Allan Prittie for help in slitting the paper to a manageable size. Prittie, who at that time had a small machine shop in the same building, later entered the slitter business as Arpeco Engineering, Ltd.

> *All they wanted me to do was print two bands of gold [to] break up that long white cigarette....It was terrible to run. Gold was a horrible color to run [in] register, but anyways I did finish it sometime before four in the morning....From then on we ran 24 hours a day, seven days a week for about a month....I got to the point that every time I would finish a roll I would put it in a cab and send it to Brampton, twenty miles away. So finally, at three in the morning, they couldn't get any more cabs to come out so the plant manager came out and he's sitting on a stool beside me on the machine while I'm running it. And he said, "What are we going to pay for this stuff anyway?" And I said, "Hey, I don't know. You've been sitting here with me, you see it takes me a whole hour to do it....I really would like to get 25 dollars an hour."* [107]

The plant manager agreed and—after a hundred rolls had been completed—offered a purchase order! Then Tomlinson found a way to make the machine run three times faster.

So rather than running at 25 dollars an hour, I'm getting 75 an hour....We made an awful lot of money. And then they said, "That's it. The stuff is on its way from South Carolina." [108]

So the day and night job was over, and he and his wife planned a week's vacation. Just as the Tomlinsons were to get in their car, the phone rang. The truck from South Carolina had rolled over, so they were shipping Tomlinson more paper.

With the profits he bought a four color Webtron and hired a pressman. Coca-Cola called with a tiny order of parking stickers. The cola company warned him he had little chance for more business because they never used stickers. Two days after delivering the one or two thousand parking stickers, Coca-Cola called again. They needed a secondary four color label concerning cyclamates.

So [they] came out with this fantastic design and I ran...25,000 of them....Before I was finished I had run twelve million....I had run six million and the government decided it didn't like the wording [so I made] another six million stickers to go on top. [109]

WE HAVE A WEEK

A few years later the president of Coca-Cola Canada called Tomlinson on a Wednesday and invited him to a two o'clock meeting on that same day. There were a number of people present in the board room. The president addressed them and told them a diet Coke was in the works and Canada was to be the first country to receive the product. It was scheduled for introduction in three months. But the problem was that Pepsi would launch its diet Pepsi to Canada within a week or two. So, the president said, we must launch a week from today!

The bottle maker had to have bottles ready, ditto the can makers, and the design artwork wasn't even complete. Tomlinson's job was a concentrate label.

On the following Wednesday night, Coca-Cola (with the help of their suppliers and two chartered 747s) scooped Pepsi Cola with their one-week launch of Diet Coke.

ONE MAN IS THE PRODUCTION LINE

Conversion processes began to make waves in the 1960s. In 1964 Mark Andy sold a three color in-line press to Buxton Skinner Printing and Stationery House in St. Louis. The Listerine labels they were printing were a simple three color label glued to the top of the package. Before the Mark Andy press, they would print one color and let it dry, print a second color and let it dry, and print the third color and let it dry. On another machine the sheets would be die cut, then a third machine would strip the excess. Finally the labels would be hand packaged in boxes for shipping.

Mark Andrews, Jr. points out:

Our press would unwind a roll of paper, it would print all three colors and die cut it, wind it up on the other end, and the pressman himself actually packed each roll in the box.[110]

PEEL 'N SEAL

Sometime in the early 1960s a phoenix arose, creating a great—albeit brief—marketplace for pressure sensitive materials. The idea was excellent. It began as a two inch by two inch square label of pressure sensitive material that closed the end of a bread loaf package. (The label later grew to 2 and a half by four inches.) "Peel" it off and take out some slices of bread, then re-"Seal" with the label. *Voila*, fresh bread throughout the life of the loaf. Bill Estridge thinks the year was 1962 or 1963.

Al Norman remembers the resealable bread label business:

I think we sold almost 2,000 Label-Aires, at that time they were only like $3000 a piece. But you can imagine what a shot in the arm that was...for Kleen-Stik....But the disappointment was that at that time there were many different types of bread wrappers. Everything from wet wax paper to different kinds of coated cellophanes, to polyethylene and polypropylene, and what everybody wanted was a universal bread end seal label....and our standard removable didn't work....Every time we [Kleen-Stik] sold a Label-Aire, Fasson got the business with their R-125. And Bill Dodge told me, when I went to work with Fasson, that we launched Fasson with that business....So we were selling the razors, but never got any of the blades.[111]

Ironically, Norman says they eventually came up with a superior universal bread end label, but by then the market was dying.

The secret to the bread-end seal was a pressure sensitive heat-applied material. Chuck Reed of Fasson recalls:

> *We had those Oliver machines on the bread-wrapping equipment that could only apply heat seals. So we needed an applicator that would put a pressure sensitive on that [which] would remove from the poly and those waxed wraps and reclose....We decided, "Okay, fine, we'd put a heat seal over the top of the pressure sensitive...so it would be peelable, re-sealable. Up to that point we had never sold more than a converter roll of material at a time.*[112]

Bakeries all over the nation were buying the application machines and the pressure sensitive labels required. It was, unfortunately, an idea ahead of its time, because no PSA supplier could make Peel 'n Seal work. Today the industry could make such a product, but in those days it was difficult to Peel and/or Seal. It often wouldn't come off without ripping the paper wrapper or it would fall off after being resealed.

> *I went around to bakery after bakery running all these tests. I would come home with half a car load of loaves of bread from our tests. We had freezers loaded, my neighbors had bread to spare.*[113]

Still the product worked well enough and often enough to sell millions of dollars worth of machines and labels, pushing the infant industry to new highs. Estridge remembers, "It was real big in those days....The first big order that I ever had was for that product and I can remember it well because it was over a $10,000 order."[114]

It didn't last long. Films and twist tops (or plastic snap-ons) superseded the Peel 'n Seal labels within—most people interviewed agreed—two years or so. But Peel 'n Seal jump started the industry. Surprisingly, it is the only PSA product application known to have become totally obsolete. Jim Wilson of MPI "knew a guy who bought one [of the T-82 Label-Aires used for Peel 'n Seal] in Illinois for 25 dollars at an auction. He thought he got a good deal. He really had it tough to keep it running because he couldn't get parts for it."[115]

"HELL" RIEGELS

A lot of water (and other fluids) went over the dam at Hellriegel Inn, still a popular eatery and bar near the Fasson plant in Painesville, Ohio. Bill Donovan of Fasson, and later Royal Label, recalls a conversation there with Burt Morgan, when Morgan was still at Fasson.

Burt said to me [Donovan], "I've got problems." I said, "What's the matter, Burt?" He said, "Well, quite honestly, some of the people surrounding me who I thought were my friends are not my friends. I've got real problems. Stan Avery is really leaning on me—says he wants me to drop completely the sale of pressure sensitive in roll form. What would that do to you?" [116]

Donovan, who had the bulk of his business in New England, said he would have to leave Fasson:

The bulk of my business is with roll label converters, and frankly, in New England I don't have too many flat bed converters that spend very much money on pressure sensitive....

The roll label people were kind of put down as entrepreneurial and hungry and very aggressive in developing new markets and new customers. They were not mired in tradition and were willing to learn something new.[117]

Hellriegel Inn was also the place that Fasson folk came to celebrate. The company had been working to produce Fascal 900, a product to compete with 3M's 3650 cast vinyl film. The samples had to be sent to the government to see if they complied with military specifications. Bev Eagon remembers that she had sent samples to Florida Testing (to see if they would pass the government tests) and announced to her boss, Les Dickard, that they would pass on all but the blue. (She immediately started to reformulate the blue.) When everything but the blue passed, just about everyone from Fasson went to Hellriegel Inn to celebrate. Eagon recalls, "That never would have happened at any other place." [118]

I LIKE COMPETITION

Gib Lewis started Lewis Label Products in September of 1964. His first order was for a thousand labels printed in blue ink on a red stock. The order cost $20.40. Lewis had paid $7,000 for his first four-inch wide, three color Mark Andy press. In his first year he grossed just under three thousand dollars. Lewis says he liked competition, "Because if we hadn't had it, I'd still be putting up the same shabby product as I did in sixty-four." [119]

Of the first five employees Lewis hired, four were still with him 27 years later! His first niche specialty was with the food industry. An early client was Sugardale Meats of Ohio. They used "a heck of a lot of labels." [120]

Lewis believes his suppliers have done a good job:

> *We've never had a problem with our suppliers. They have always been handy and wanted to help us through good times and bad times....I think that's what I like about this industry. Here the competition gets a little heavy from time to time—there's more of us. There's still the good friendly competition, sharing of ideas....We can do almost as much business for our competitors as we do for somebody in direct customer sales.*[121]

COMPUTER LABELS/DATA TRANSFER

If you remember the early days of computers you will recall the hue and cry of, "Paper is doomed. The electronic office is coming." What could have been said was, "Computers (and the copy machine) will change the way companies use paper. In ever-increasing quantities."

One application was the computer-generated mailing label. Pressure sensitive adhesive material was ideal for the mailing label. Alan Campbell, who has worked for Avery/Fasson for 29 years, began working with business forms converters in developing EDP products. "That included working with the people that just produced blank pin-fed labels and also companies like Moore Business Forms... [incorporating] pressure sensitive labels and forms into collated sets." [122] He further points out that information processing represents about 50 percents of the total market. "It's the printing of variable information. Getting variable information out of the computer or some other place onto a label."[123]

APPLIED COSTS

Computer labels made a big change in Ever Ready's product mix in the early to mid-1960s. Harry Stillman said that before then most address labels were water gum and...

> *People like utilities and banks were running hundreds of thousands of them and had no method [of applying them]. They had to tear them apart at the perforation like a postage stamp and then find a way to lick them and stick them. [There was some] machinery that was used to apply them that was semi-automatic, and then dispensers came out and so forth. [But applying the labels became] labor intensive and applied costs became a factor.*[124]

Applied costs means the total cost of the label and the cost of putting the label on a product. Gerry Gartner of GarDoc says his company always does an applied cost analysis. For example, the main competitor of pressure sensitive labels, says Gartner, is therimage labels. With an applied cost analysis, "sometimes therimage, sometimes PS would win. And because we were relatively new...there was no way you could compete and win if you were just comparing label price. So you had to look at the true applied cost." [125] GarDoc would go into the client with a beautiful label <u>and</u> a full analysis:

> *[And we'd say], "Isn't this a beautiful label for $10 a thousand. They'd say, "Hey, I can get one that looks just as good for $2.50 a thousand." Then we'd say, "Yeah, but it costs eight dollars a thousand to put that one on and it only costs fifty cents a thousand to put ours on." It's a tough sale because companies are used to looking at the sales from the finished point of view. But that's all changed.*[126]

EASY MONEY

With improved economics, the pressure sensitive salesman of the sixties made major inroads into larger and larger companies, increasing the volume of PSARL materials dramatically. As Duane Hillmer of Epson Hillmer Graphics said, "If you did anything at all right, why you were better than a survivor. You made some money." [127]

Stillman suggests the EDP label made a vast impact on the industry because "once a label had pin feed holes in it, it became in their eyes [the business forms industry] a business form, no longer a label product."[128]

Now the thousands of salespeople from Moore, Uarco, Standard Register and other giant companies were selling pressure sensitive labels:

> *All of these companies subcontracted those products initially to people like us [Ever Ready] or Avery. Moore [the largest forms manufacturer] went to Avery...Uarco came to us for a while, and ultimately Standard Register came to us about ten years after they had been in the field. But they were doing at that time two and a half million dollars 25 years ago.*[129]

Uarco may have been the first forms company to take their pressure sensitive business in house. Even with most of the larger forms companies doing their own printing, the pin feed label application business for PSARL has continued to grow. There are specialized firms which primarily produce EDP labels, such as Data Label, Metro Label, Continental Forms and Label America.

NICHE PLAYERS FORM AN INDUSTRY

Jim Kyte of Coated Products says the mid-sixties was when pressure sensitive labels changed from a hit or miss niche player to an industry. Although Dow Chemical had introduced silicone for producing release liners, it was the introduction of General Electric as a supplier of silicone that changed the picture, says Kyte. That and more competitive pricing.

> *[Pricing] really opened the market up. [Before] the pressure sensitive label stock base material was twice what a heat seal or gum label stock would be. All of a sudden it's down there nip-and-tuck. And then they started to go with high speed automatic labeling devices. Then the quality of the product improved to a point where it was cost effective to substitute pressure sensitive for the gummed or heat seal....It exploded at that time....Removable was a big thing at that time because that was something that you couldn't do with the heat seal or a gummed label. [The PSARL industry] started to explode in the early sixties when the prices came down and then really took off in the late sixties when everybody's volume picked up.*[130]

Kyte says Kromekote (and other cast coated papers) was a mistake, "a mistake that grew into a big business." [131] Kromekote was an excellent printing paper for graphic arts printing on letterpress or offset, but its use in flexographic printing was shaky at first because "they [Champion Paper] didn't make it specifically for flexographic inks. So then [the ink suppliers] developed something that would work. Today it's a great market." [132]

SPRING HOUSECLEANING

Everyone knew that adhesives would not stick to silicone:

> *That was a given. When we were making the old release coatings—and Avery did the same thing, Kleen-Stik did the same thing—[we] would put down a varnish and put down an oil—a heavy oil—and bloom it out....The gimmick was you can't bloom enough of it to detacify the adhesive, but there has to be enough to get it off. That was the dichotomy these chemists went through for years. It was a son-of-a-gun because we had to have two release coatings—one for summer and one for winter. If you didn't clean house in the spring you were in big trouble.*[133]

OPERATOR-DRIVEN

In looking back at his many decades in the business, Jim Kyte says he was struck with the fact that even today with huge companies doing billions of dollars in business, the label producers are still operator-driven:

> *They kind of like got into the development area as an adversary. For good reason. If his output was slowed down, it was the fault of the label stock, it wasn't his fault that he was out drunk last night or something like that....The label printer, and some of them are pretty darn big right now, are still job shops. A job shop is primarily operator-driven. I ran into it yesterday. I was amazed that after all these years that most of these companies are still operator-driven.*[134]

1. BUY PRESS. 2. START COMPANY

"It's so much fun and so easy," [135] was how Don McDaniel described selling pressure sensitive labels. And for good reason:

The company I was working for had been in business already ninety-some years prior to my being there. Although they liked the pressure sensitive business and saw a future in it, they weren't quite as excited about it [as I was] because they had been doing pretty well with the business they had been doing all those years.[136]

Salem offered letterpress and offset as well as flexo, and had just purchased a gravure press as well, planning to enter the printed can label and film market. [137] Each printing process produced products which competed with the other for business and the salesmen's time. McDaniel wanted to concentrate on flexographically-printed pressure sensitive labels which he believed "had a real future."[138]

So I just went out there and went off on my own and bought a used Mark Andy....The one I bought was six or seven years old, but had really never been run much. I paid a thousand dollars for it. [The press came from Firestone Tire & Rubber.] They were going to print their own builder numbers.[139]

In those days a person actually built a tire. A builder number was a quality control measure, inserted before the final layer of rubber was applied and vulcanized. The customer never saw the number but if the tire failed or wore out too quickly under warranty, the tire came back to the tire company which could easily discover the builder number. Naturally the tire builder didn't want his number showing up too often.

But Firestone discovered they could buy their maker numbers more cheaply from Excelsior, which McDaniel says, "was king of the tire labels and tire builders numbers and tire tapes in those days."[140]

MPI Label System's first customer was the Genie Company of Alliance, Ohio, which made television antennae rotors and fractional horsepower motors for the little fans that were used inside Kelvinator refrigerators. The second success came from the meat packing industry.

Tom Wilson, who worked with McDaniel at Salem Label and later joined him at MPI, recalled one of the really big orders:

It was printing these Champion spark plug labels. You know it boggled the mind some time when you're running something like that...an order for 82 million. You print those things and they're coming off the press three wide and a girl's catching them....We would run that two shifts a day....We'd see one shift would do 800,000 or 900,000 or one million a day. We'd see if we could do better the next day and both shifts would do that. Man, we would run those things.[141]

BIG TIME

While attending an Ed Vandenberg seminar, McDaniel ordered a six inch, two color press, but it wouldn't be ready for a few months, and MPI needed a press right away.

So I called Jim Brown because Jim was making the International Presses at National Converters, which is a press that Ed Vandenberg had developed when he was there. And Jim didn't have any new presses to sell, but he said his business wasn't all that great, "I'll sell you one of the presses off the floor of my label company if you want it." [142]

The machine involved, a six inch, three color International, was doing one of the last runs of Peel 'n Seal labels on a Saturday morning. McDaniel and two of his employees drove to Cincinnati on Friday and stayed the night. After the Peel 'n Seal had been run, the press was cleaned, loaded on McDaniel's pickup and hauled back to Sebring, Ohio. "Not long after this our new six inch, four color Vandenberg arrived....We were big time. We had as many presses as anybody." [143]

MPI currently has many more plants than they had presses then.

SLIPPERY CHIQUITA

In 1968, when Dick Pearson was running Avery Label, the infamous banana incident occurred. Mentioned by many of the people interviewed for the book, the story goes like this:

Chuck Reed was sent to South America because the Chiquita banana labels kept falling off the bananas. The labels tested out okay, so Reed asked to see the plant. Immediately he could see what had gone wrong. Bananas often come with resident tarantulas, so United Fruit had its workers dip them into a bath to kill the tarantulas. Then the labels were applied. Since the bananas were slippery, the labels curled and fell off.

Avery gave United Fruit a credit of $110,000. [144] The bananas were dried and the labels stuck.

A footnote to the story is that a Florida company devised a way to put huge plastic bags over the developing bananas, thus keeping the tarantulas away and the label problem disappeared.[145]

STORMIN' NORMAN

Al Norman, a 1952 New York University graduate, served during the Korean War and joined National Starch and Chemical Company in 1954 doing technical service for their adhesives operation. He wound up ten years later as manager of their resin specialties group which involved all of the company's industrial adhesives and coatings, including pressure sensitive coatings. In 1964 he was asked to go to Kleen-Stik, which National Starch had acquired in 1961. He was to be the technical director. It was at this time, Norman says, he "began to learn very quickly how little I really knew about the intricacies of the pressure sensitive converting business."[146]

He went to Plainfield, New Jersey, and a small group reported to him:

> *[We were responsible for] quality assurance, which we had very little of, and I had to build technical service, which we had none of, and I had to create market development, which we didn't understand, and I had to do it myself, and a plant that didn't know why it was there and we had to figure out why.* [147]

Norman echoes the other Kleen-Stik alumni who rued the loss of National's Jim Dillon and the "orphaning" of Kleen-Stik at National Starch.

> *It was not profitable. Lots of potential but not profitable, very fine and innovative marketing people with good ideas but some of the worst manufacturing equipment and practices you've ever seen in an old decrepit plant. Along came a gentleman by the name of Garrison Brinton, who at one time had been a DuPont executive....*[148]

Norman was "sold" to Compac Corporation in 1968 as part of the National Starch sale of Kleen-Stik. Compac was the work of three men, Brinton, John Dickenson and Vic Pollock. A new boss—"to say that the chemistry was poor between us was a gross understatement"[149]—caused Norman to find employment with Fasson in 1970. He worked for Fasson until April of 1979.

WHO YOU GONNA CALL?

The careful reader of this book will immediately guess where Norman would go after leaving Fasson. It's elementary. MACtac, of course. And Norman has remained there ever since. "There were only so many heavy hitters in the early days of this business and if you were any good at what you did, people knew you." [150]

Norman is proud of being the chair of the TLMI committee that developed the first high speed release tester and test procedure in 1968. It came in under budget and was about three times as fast as anything available at the time. "[Then we could talk] a common language about this crazy thing called release."[151]

NO MOOSIC, PLEASE

John Orlando stayed with Simon Adhesives after Litton Industries acquired them in 1960 or 1961. But the proposed move to Moosic, Pennsylvania was not for him. He left on a Friday and on Monday was working for Jim Kyte of Coated Products. For one year; it was a 126 mile-a-day commute.

Then Orlando went to Kleen-Stik where he handled just four accounts. But each was at least a million dollar account in 1967. He rarely left his home, doing all his work by telephone. "It was awfully boring. But that's what I did."[152]

He handled Kimball, Ever Ready Label in Belleville, New Jersey, Alan Hollaender in the Bronx and Alan Hollaender in Cleveland. He left within a year. "They were paying me more money in those days than I had ever seen in my life. And to take that kind of money and do nothing, I just couldn't do it."[153]

Then Orlando joined Pressure Stick, the company started by Mark Simon, Sander Simon's brother. Mark Simon had been out of the business for a number of years before starting Pressure Stick, which was later sold to Rockwood Industries. Then Orlando joined Al Sobel's Coronet Paper and sold—by phone—millions of pressure sensitive EKG mounts.

> *We went cross country. Just a fabulous thing. And Al let me run [the business] like it was my own. [Al Sobel would] take a piece, what people would term junk and make it do things it had never been meant to do before. He devised a way of producing [PSA] Braille labels—Mrs. Sobel was very active with the organization for the blind.*[154]

John Orlando was wooed away from Coronet by Joe Sanski of Patton, where he acted as a consultant for the next three years. Orlando retired on October 14, 1993, but states that he is still on call.

1000% BETTER

Richard Rosemann believed it was 1968 when he first came out with a chrome-plated die. In some cases this increased the life of the die by a thousand percent!

Rich Mark, the Seattle company founded by Mark Donner, had to create an oval label that had white pigmented ink on foil. "He'd get 50,000 revolutions and it would quit. And he had an order for a million labels."[155]

So Rosemann found a St. Louis chromeplater and produced a die that ran the rest of the million labels and was still going!

Strangely enough, in his interview only months before he died, Rosemann said his standard grade die was still using the same non-hardened steel he had been using since 1969.

"Tyvek was a nasty material to cut," Rosemann recalled:[156]

> *I would like to say that I learned to cut it, but I don't think that was the case. [DuPont made some] changes in the material. Mylar got to a point that we could fine tune dies well enough to cut Mylar as well as paper....For a long time my dies were general purpose and they would cut Mylar, but I wouldn't guarantee they would unless [the customer] specified it. If they specified it, then I tested it on Mylar. To test every die on Mylar would have been cost prohibitive.*[157]

Rosemann says he regretted not expanding rapidly enough in the early days "to keep it all to myself."[158] Here's what he says about his business success:

> *One of the things [to which]I contribute most of my success to is...when I started making dies, money was not the prime object. It was making that die....I had to get enough money to keep on operating. But from there on I didn't care...and I'm doing pretty much all right today*[159]

He believed in training his own machinists, because his processes were "counter to good tool and die making rules." [160] So he would hire only inexperienced workers and train them himself.

STAR OVER THE PHONE

One day Lee Paul (Tape and Label Engineering) called Rosemann and said he wanted to order a star. Rosemann said he needed a sample and Paul answered, "Why don't you shut up and listen to me?" "He wanted a six-pointed star...a star of David. And he told me exactly how to make it." [161]

Although he credits his customers with coming up with challenging ideas, Rosemann is proud of creating the hydraulic monitoring of the die pressure.

> *I wanted to take the human factor out of it when the dies were tested. I wanted to be able to know that they were tested...under a certain pressure. Not a pressure a guy would crank to make it work [off the press]....When I saw how well it worked as a quality control...I was able to get it on the presses themselves.*[162]

A YEAR AND A HALF FOR FITCHBURG

Bev Eagon left Fasson in 1969 and went with Fitchburg. Litton had purchased the Simon Adhesives operation and moved it from Long Island to Scranton/Moosic, Pennsylvania.

> *One thing about Fitchburg, I'll say they did buy good equipment. Unlike MACtac. They went out and they would see what Fasson had and they would say, "We want one, just like Fasson's." It [Fitchburg] had all the potential but it didn't have the leadership that they needed or they could have been right out there in the front. Anyway [after] a year and a half of that, I said this place isn't for me. So that was when I went to MACtac.*[163]

Eagon's evaluation of MACtac will be a highlight of the 1970s.

THE MAKING OF MILLIONAIRES

Maynard Louis had worked for Dennison for eight years before he and his next door Chicago neighbor, Les Ordman, decided to enter the label business on their own. Louis said he "took a second on the home and borrowed all I could from the family."[164] Both Les and Maynard were college-trained "peddlers," Louis says, and they started Lord Label in 1965. They purchased a multi-flex from Webtron and immediately specialized in the bakery business, going for automatic applications.

Their competition at the time was Avery, Dennison, and Steigerwald, so Lord Label had to choose marketing niches carefully. Louis says they worked 100 hours a week in their first year, 1965. The second year "we cut it to 95; we didn't come in on Sunday."[165]

Bruce Motter, who worked for Fasson, MACtac, Chemtrol and Technicote, used to call on Lord Label. "This wasn't too many years ago. They would sit in a [small] office... face to face. They had a cardboard box, one 'in' and one 'out'." [166] Both principals were salesmen, and Les supervised production while Louis did the financial work. But Louis says:

> *We worked on the machines. There wasn't anything we couldn't do. We were not good pressmen. But we did work on the machines. We knew the business from A to Z. [Our employees] respected us because we worked with our hands and our heads....We conserved capital, we put everything back into growth, we didn't take exorbitant salaries, we always knew where we were at.*[167]

One of their first orders was from Jewel Tea, still a Lord customer today. "Kraft," says Louis, "has been wonderful to us...and Armour was good for us also." [168]

Louis credits a key employee named Gene Stokes for craftsmanship that allowed Lord to do "things that maybe the others couldn't do" [169] for Kraft labels. Stokes, hired as a pressman, became a Lord plant manager.

Lord Label had hired Phil Henning in August of 1965 to operate their first narrow web flexo press. Phil managed the continuing growth of the Chicago manufacturing operation for the next nineteen years. As Lord Label grew, Henning groomed his oldest son, Philip III, to start up the Dallas manufacturing operation in 1976.

THE DYNAMIC DYNAMO

Burt Morgan offered ninety-day credit for Lord Label in the early years and is remembered with great affection by both partners. "His giving us ninety-day terms did more to help our business than anything," [170] said Louis, who also credits Richard Rosemann of Roto-Die and David Mages of Webtron for their contributions to Lord.

> *Burt improved the equipment and cut the prices literally in half—not figuratively, literally—from 26 to 13 cents a thousand square inches. And that made the pressure sensitive industry competitive with external glue-applied labels.*[171]

Les Ordman is especially proud of three Lord Label innovations. One was to produce automatic applicators in-house for specific customer needs. Originally they simply planned to buy them from Label-Aire. Why re-invent the wheel, they thought, but they found one size doesn't fit all. Rather than modify they decided to build. Ordman said:

> *And we came up with a machine of our own design which, instead of blowing the label on, we tapped it on....The label went out on the grid and the grid, instead of having a backing that blew it down...we held it on the grid and put the grid on a piston and carried it to the container. Building machinery was one of the best moves we ever made.*[172]

The second was the creation of regional plants:

> *We recognized early in the game that you're only viable within roughly a 500 mile radius of the plant. When you get beyond that there are too many good converters out there....Now the sincerest form of flattery is to be copied and since we've done that MPI has done it and so has CCL.*[173]

The third innovation Ordman recalls was the realization that the dairy industry would shift to plastic containers. Plastic containers were a big boon to all pressure sensitive label producers because water gum labels will not stick to them.

The conversion to plastic containers began in the southeast and Lord Label, with its multiple plants, was able to service them. The first milk container label was bullet-shaped and placed in the handle area. This was fine with the dairy people, but the supermarkets hated it. They wanted lots more on the label—logos and selling copy and later, UPC codes.

The problem is that if the label is positioned elsewhere (beyond the protection of the handle) it must be protected against scuffing. As Ordman says, "There isn't a more hostile environment than a dairy. It's wet and cold and the product rides to market in a container that is not tight, so it shifts."[174]

Lord developed an inexpensive lamination which solved the problem. "It was our leadership role in solving problems of the dairy industry that propelled us to a very strong position in the label field," Ordman said.[175]

THE MILLION DOLLAR EGG

Lord Label attracted a lot of industry attention when the company developed a recyclable polypropylene label in the mid-1980s. As Ordman says, there was never an attempt to keep the label a secret,

> *We were very open about it because we knew if we created a market for this we'd never be able to supply it. It was for everybody. We wanted folks to start thinking about it. But it carried about a ten to fifteen percent premium...and folks said, "Well, that's a terrific idea and maybe we'll do it when everybody else does it."*[176]

So Lord had produced a superior product that is stronger, more moisture resistant and recyclable as well, but one that few customers will pay more to use. Mobil Oil and Dean Foods are the exceptions, but environmental concern may save what Ordman called their "million dollar egg," adding that Germany has a new law requiring that all packagers must have one-third of their consumer materials recycled, so perhaps the egg will prove golden in a few years.[177]

Les Ordman credits Lord Label employees for playing a major role in the company's success.

> *Mostly unorthodox, entrepreneurial, ambitious, some limited education, but just had other skills. We were blessed, really blessed with good solid people....When these people threw their hats in the ring with us, we told them straight out, we'll build it up, get to the point where it'll be attractive, and then we'll sell it....They knew from the get-go...the game plan.*[178]

When Lord Label was sold in 1989, five of the company's employees became millionaires in addition to the two founders. Ordman says:

> *One of the greatest satisfactions of the whole career is to look back and say, "Hey, those folks who are now our friends that came with us during that time are set for the rest of their lives."*[179]

The only key employee planning to leave is Maynard Louis, who says, "It's too tame after being at risk." [180] So he has invested in two other companies and plans to be very active in each.

ORGANIZING THE INDUSTRY

In June of 1933, as a result of the National Industrial Recovery Act, the NRA (National Recovery Administration) required manufacturers to form trade associations. Each trade association would then deal with the NRA on behalf of its members. TMI, the Tag Manufacturers Institute, began with twenty members.

> *From the spring of 1933 until the spring of 1935, every trade association in every industry concerned itself almost exclusively with the creation of, and learning to comply with, an NRA "Code of Fair Competition" for the industry it represented.*[181]

TAG MANUFACTURERS INSTITUTE

The tag manufacturers cooperated by filing their monthly sales figures with the Baxter Association, which disseminated the industry figures to all association members. "Although they didn't tell how they did certain jobs they were able to do less throat cutting in their pricing so they could expand better." [182] This helped stabilize prices within the industry. "If you were found to have gone off the list and had not reported it, you were subject to a financial penalty...levied by the Baxter Association." [183]

In 1949, as a result of the statistical program, the government intervened and the manufacturers were taken to court for price fixing. The case got as far as the Supreme Court, but the court refused to hear it, so no manufacturer was fined. It was a landmark decision which "reconciled the government's efforts to enforce the anti-trust laws with practical economics." [184]

THE ROLL LABEL MANUFACTURERS ASSOCIATION

The Tag Manufacturers Institute was not enough for those labelers who used pressure sensitive materials and a few individuals joined together to start the Roll Label Manufacturers Association. Bill Muir of Grand Rapids

Label recalls most of the members of the Roll Label Manufacturers Association, which he believes started in the early 1950s:

Grand Rapids
Package Products
Ever Ready
Steigerwald
Alan Hollaender
Fairbane (San Francisco)
Miller and Miller (Atlanta)
Duane Hillmer

"Plus one or two other members and they met two or three times a year for a number of years. [It] wasn't strictly pressure sensitive, but almost everybody was in pressure sensitive." [185] The Roll Label Manufacturers Association and the Tag Manufacturers Institute competed for members during this time.

TLMI

In 1962 the Roll Label Manufacturers Association and the Tag Manufacturers Institute merged. The organization was renamed the Tag and Label Manufacturers Institute (TLMI). Twelve pressure sensitive label manufacturers were the first members of the new Label Division. John Torrey (Avery Label) was elected Chairman. By the way, his 1962 prediction that the pressure sensitive label industry would reach $100 million by 1970 was woefully low! The actual number was closer to $240 million.

In 1966 the Institute formed an Associate Membership Division, which was open to suppliers to the industry. Fifteen charter members joined this division.

In 1991 TLMI became a self-managed association, headquartered in Iowa City, Iowa. The institute is international in scope, and in 1993 its membership totaled over 220 member companies. TLMI's mission is...

> *to be the leading trade association of the tag and label industry by providing a forum for the exchange of ideas, information, education, and the establishment of appropriate industry standards.*[186]

To accomplish this mission TLMI conducts statistical studies such as ratio studies, wage and labor reports and the roll stock report. It sponsors membership meetings and technical conferences and conducts an annual awards competition that stimulates innovative printing, recognizes technical achievement and serves as evidence of the maturity of the industry. TLMI also provides marketing and public relations benefits and publishes the *Glossary of Terms for Pressure Sensitive Label*, and standards manuals for both tags and labels. The organization informs members about important events and issues through its bimonthly newsletter, the *Illuminator*.

Things were slightly different in the early days. Herb Smith remembers, "When I first joined everybody played golf and we tried to limit members to golfers only. Then came tennis players and finally fishermen. The associations and meeting content was invaluable." [187]

SECRETS

Even then, secrecy was already a religion in the field. Some corporations would not even let salesmen into the factory. "There were people that had such innovative ideas that before you could go into their plant you used to have to sign a non-disclosure agreement or they would strip all the mechanism off the press before I would go in to do repair work," says Leon Beaudoin. [188] That attitude still exists to some extent today, but in addition to supplier training programs and newsletters, much of the secrecy has been dissipated through the early Baxter Association, and later the work of TLMI.

A notable contribution to these changes has been the TLMI Management Ratio Study. In 1973 Gene Singer began to collect and analyze confidential key accounting data submitted to TLMI by member firms. This series of statistical surveys, named for many years the "Singer Ratio Report," "gave certain goals and comparisons which were perfectly ethical and perfectly legal to do. He gave us a sense of purpose." [189]

In 1986 TLMI issued the Best Managed Company Award to three companies, divided according to the volume of sales of tags and labels, with the most favorable financial ratios indicating excellence in business management. The study was remnamed the TLMI Management Ratio Study, and in 1992, the award was renamed the TLMI Eugene Singer Award, in honor of the study's initiator.

Winning companies since the inception of the Singer Award have included:

Aladdin Label Inc., 1986
API Graphics, Inc., 1986, 1991
Artcraft Converters, Inc., 1992
Atlas Tag Co. of Canada, 1988, 1993
CL & D Graphics, 1990
Graphic Solutions, 1992
L & E Packaging, Inc., 1987, 1990, 1991
Label America, 1993
Labelgraphics, Inc., 1993
Lancer Label Division of Porter Chadburn, 1987-1992
Mail-Well Envelope Co., 1986, 1987
Menasha Corp./Mid-America Div., 1989
Model Label, Inc., 1988
New Jersey Packaging, 1989
Tapemark Company, 1990, 1992

THE SINGER REPORT

Eugene Singer, who joined Avery in 1955, began market research studies in the late 1950s when Avery was expanding at 25 to 45 percent a year. With eight or nine new sales territories every year, Avery felt it was time to look at what the industry had in common, "particulary in such areas as technology, press development, legal aspects. It had nothing to do with competition in the strict sense of the word." [190] Singer's job at Avery was to develop "the management ratios trying to find some kind of common information of both marketing and technical."[191]

Singer then transfered the concept to TLMI in 1972. TLMI had been doing management ratio, although only informally. "I virtually had to sign my name in blood to the trade association that I would not reveal anything that I'd learned through doing any ratios or any financial analysis for the companies."[192]

Singer also was using the Avery computers to run the TLMI ratios! He had a time share access to it and "nobody had access to the numbers." [193] Probably most telling is Singer's statement that, "we could have never succeeded with the ratios over a period of almost twenty years unless they believed what I said, [that] nobody ever sees these numbers."[194]

The management ratios work. Singer tells of...

> *one guy, I can't remember his name....He jumped right up in the meeting one time and said, "Last year was the first time I ever participated and I got those numbers. I realized what I was doing wrong and I went back and I sold off half of my equipment and I'm making more money than I've ever made before."* [195]

The Singer Report continues to be produced by TLMI, showing companies different insights into their operation, "sharing, but at the same time not revealing. The whole [concept of the] management ratio is that. Sharing, but not revealing."[196]

BOB KLAS ON GENE SINGER

"He's a remarkable guy," says Tapemark's Bob Klas, Sr., "and I think our industry owes a lot to Gene Singer."[197]

Klas noted that the early entrepreneurs were ingenious, but unsophisticated. But as time went by,

> *There came a time when our industry began to be recognized as an industry and...various suppliers began to pay us serious attention....Gene Singer played a very important part because each year he used to furnish a thing called the Singer Report. Voluntarily people used to submit confidential key data. And his statistical survey of what we were doing gave certain goals and comparisons which were perfectly ethical and prefectly legal to do. He gave us a sense of purpose....We actually had guys in the industry...who would have been better off to liquidate...and draw a passbook savings interest rate.*[198]

CONVERTER OF THE YEAR

In 1987 TLMI cosponsored the first Converter of the Year award with *Package Printing* magazine. Award winners include:

1987	Don McDaniel, MPI Label Systems
1988	Darrell Dochstader, GarDoc, Inc.
1989	Jim English, Kalamazoo Label (now KAL GRAFX)
1990	Pat Patrick, Label America
1991	Dick Schwartz, Aladdin Label
1992	George Noah, Lewis Label Products
1993	John O'Brien, Porter Chadburn, Inc.

LABELEXPO USA

Another function of TLMI is the organization and co-sponsorship of LABELEXPO USA. This converting industry trade show began in 1989, featuring the most extensive collection of tag and label equipment and supplies found under one roof. Tag and label manufacturers alike praise the concept, and suppliers feel they have the best forum available. "No other show can offer us that value for our promotional dollar," said Dale Bunnell of Mark Andy, Inc.

INTERNATIONAL COOPERATION

Don Buchta sums up TLMI well:

> *Being a member of TLMI brought me closer to the whole industry and it's a good vehicle for me to get into contact with other producers of various products that I wasn't aware of before we became a member of TLMI. I just felt that I not only helped broaden horizons but...we cut standards for the industry which I think was very important at the time.*[199]

Buchta, as President of TLMI in 1983 and 1984, helped the organization contact and network with FINAT, its European equivalent (Fédération Internationale de Fabricants et Transformateurs d'Adhésifs et Thermocollants sur Papiers et Autres Supports). "It's broadened our horizons, but again made us aware of other applications and uses that we weren't aware of here in the states."[200]

Since 1976 TLMI members include international firms as well as North American companies. (The first Canadian firm to join was Labelcraft, the first Mexican firm was Industrias TUK, both in 1975.)

HOW NARROW IS NARROW?

George Parisi, Executive Director of FTA said that "narrow web is the fastest growing segment of the flexo industry."[201] But, when it comes to defining what is narrow:

> *We have that problem among our own board members.... Narrow web could be anywhere up to, let's say 34 inches, but the [pressure sensitive] label itself, we look at sixteen inches.* [202]

While the FTA defines narrow as sixteen inches or less, Parisi added, "Different people put different dimensions on it...but you could be talking about 34 inches."[203]

Parisi pointed out that narrow web is now printing folding cartons, for example, the boxes for playing cards, a toothpaste tube box. Other unusual applications include gold stamping:

> *There's one company—I think it's out in Nebraska—Riley Advertising. They created a sort of little niche market in [gold stamped] advertising products....point of purchase... there are so many opportunities.*[204]

FTA

The Flexographic Technical Association (FTA)was first organized in the late 1950s because of the "need for some sort of organization to provide technical interchange of information for the overall advancement of flexographic printing." [205] The organizational meeting was held in New York on September 26, 1958, with Karl L. Weik elected first president. Julian Ross was the organization's first executive secretary, followed in 1968 by George Parisi, the current Executive Director. The 125 charter founders have grown to more than a thousand FTA members.[206]

FTA is heavily oriented toward the roll label printer. Mark Andrews, Jr., first learned about FTA in 1963 when he attended a workshop on the basics of flexography sponsored by the organization. "The more I got involved with that kind of environment the more I found that people were willing and anxious to share information for the good of the industry.[207]

According to Andrews, Jr., the evening sessions "would go into the wee hours of the morning when people would just tell about the jobs they ran last week, the problems of it, and what worked well, what didn't work." [208]

The FTA awards competitions date to the beginning of the organization. Andrews pointed out that the competition winners, which are shown in workshops all over the country, "are a tremendous incentive for people who wanted not only to improve their own quality, but to get recognition for having done really good quality." [209]

The FTA textbook has become "the most definitive work on the Flexographic printing process." [210] It has been translated into Spanish, as well, an indicator of the large role the FTA plays in international flexographic activities.

In 1976 the FTA Foundation was established. The Foundation's Endowment Fund was instrumental in creating in-plant training aids which benefit the entire industry. Several scholarships are underwritten by the Foundation. FTA is another example of the incremental growth of the PSARL industry.

FROM CANNERY ROW TO MODULAR MACHINES, PART I

Bob Scheerer's experiences are so varied and interesting he needs to be the subject of a sidebar for the 1960s and the 1980s.

Scheerer was born in Seattle, went to the University of Washington to earn a degree in engineering. He served in the military, becoming a ninety-day wonder and invented a rifle-cleaning machine while stationed in Massachusetts. Prior to that he and his father had built a cannery in Alaska and another in Washington state. Working for his father was not what he wanted to do, so he went to work for St. Regis Paper. He remembers St. Regis fondly, because they would assign their new hires to a company retiree:

When I joined the company I was assigned to a fellow named Art Wood...who had been the operations manager for the west coast operation. So you were getting training under people who were irreplaceable with all that experience.[211]

After eight years on the west coast, becoming manager of packaging systems, Scheerer was moved to West Nyack, just north of New York City. About 1964 he realized he "couldn't afford to live in New York."[212]

Another St. Regis alumnae had joined Avery Label, so Scheerer accepted a position and moved to the Los Angeles area to work at the Avery plant in Monrovia, where he worked on automatic labeling systems.

Since his contract with Avery had expired, Scheerer resigned and planned on returning to St. Regis, when Dennison contacted and hired him.

At that time Dennison had six plants, many of which were tag operations. They made their own adhesives and release liners. In the late 1960s, sales of the industrial pressure sensitive materials were only about $600,000 annually, approximately four percent of Dennison's total sales. Scheerer was to bring Dennison into the industrial label business.

CHRONOLOGY OF APPLICATIONS, 1960-1970

1961	Bread Labels/Peel 'n Seal [213]
1964	Demise of Bread Labels
1965	Sock Brand/Textile Rider Labels[214]
1966	Banana/Fruit Labels
	Drum Labels
	Tire/Battery Labels
	Pharmaceutical Labels
	Multi-part Address/Prescription Labels

CHRONOLOGY OF INVENTIONS/ INNOVATIONS, 1960-1970

1962	Webtron "User Friendly" Roll Label Press
1965	Doctor Blade[215]
	Glass modified Flexographic Plates
	First Flexographic Printer/Process Seminar
1968	Chrome Plated Rotary Dies
	Start of Flexographic Process Color PSA Roll Labels
1969	Light-Fast Inks
	U.V. Coatings

THE MAN WHO WOULDN'T QUIT, PART FOUR

After Compac Corporation purchased the former Kleen-Stik operation from National Starch and Chemical, Harold Scherer remained. There was a California plant at the time, and in the mid-sixties he was in management, purchasing and labor contracts for the west coast operation. In 1965...

> *We were right in the middle of the Watts Riots...and there was a man killed on our corner so I closed the plant up and sent everybody home. We didn't return for three days....Not long after...they decided to shut down the manufacturing operation and manufacture everything in the east. And then they even shut down the manufacturing in Chicago. And we became like warehouses.*[216]

In 1968 Scherer was ready to leave the firm but talked to George Collons first. Once more, Scherer was persuaded to stay because the company didn't want to lose him. His ultimate decision was to stay, but to go into sales. Scherer had many contacts because he had to go outside a lot to try to solve customer problems, "and believe me, in those days we had some big ones."[217]

[1]Muir interview, p. 5.

[2]Mark Andrews, Sr., successfully ran Kleen-Stik against Fasson materials.

[3]Beck interview, p. 3.

[4]*Ibid.*, p. 14.

[5]Dick Pearson interview with Bill Klein, p. 4.

[6]*Ibid.*, p. 2. Fasson's business was $2.3 million in 1960.

[7]*Ibid.*, p. 3, 4.

[8]*Ibid.*, p. 4, 5.

[9]*Ibid.*, p. 12.

[10]*Ibid.*, p. 13.

[11]*Ibid.*, p. 18.

[12]Don Foukal interview with Bill Klein, p.3.

[13]*Ibid.*, p. 5.

[14]*Ibid.*, p. 5, 6.

[15]*Ibid.*, p. 7.

[16]*Ibid.*

[17]Kyte interview, p. 14.

[18]*Ibid.*, p. 15.

[19]*Ibid.*, p. 18.

[20]Schwartz interview, p. 13, 14, 15, 16.

[21]Norman manuscript, p. 2.

[22]Collons interview, p. 25.

[23]Schwartz interview, p. 16.

[24]*Ibid.*, p. 17.

[25]Lasinsky interview, p. 6.

[26]Schwartz interview, p. 21.

[27]*Ibid.*, p. 22.

[28]Jim Kyte noted that Coated Products had Crocker Burbank produce Kraft super calendering as a first around the same time and Al Norman dates the Crocker Burbank SCK at about 1965.

[29]John M. Questel interview with Bill Klein, p. 2.

[30]John M. Questel questionnaire faxed to Bill Klein, November 9, 1992, p. 1.

[31]*Ibid.*, p. 2.

[32]*Ibid.*

[33]Bev Eagon interview with Bill Klein, p. 1.

[34]*Ibid.*, p. 3.

[35]Clark, The First Fifty Years, p. 80.

[36]Eagon interview, p. 5.

[37]*Ibid.*, p. 5.

[38]*Ibid.*, p. 10.

[39]Mark Wert interview with Bill Klein, p. 1.

[40]*Ibid.* p. 1, 2.

[41]*Ibid.*, p. 28.

[42]*Ibid.*, p. 3.

[43]*Ibid.*, p. 7.

[44]*Ibid.*, p. 11, 14.

[45]*Ibid.*, p. 19.

[46]*Ibid.*, p. 22.

[47]Egan interview, p. 14, 15.

[48]Andy Beck interview with Bill Klein, p. 6.

[49]Al Sobel interview, p. 11.

[50]*Ibid.*, p. 7.

[51]Duane Polkinghorne interview with Bill Klein, p. 1.

[52]Carlson interview, p. 8.

[53]Polkinghorne interview, p. 2.

[54]Leon Beaudoin interview with Bill Klein, p. 4

[55]*Ibid.*, p. 3.

[56]*Ibid.*, p. 4.

[57]Morgan interview, p. 24, 25.

[58]*Ibid.*

[59]Beaudoin interview, p. 5.

[60]*Ibid.*, p. 7.

[61]Polkinghorne interview, p. 6, 7.

[62]*Ibid.*, p. 13, 14.

[63]Reed interview, p. 24.

[64]Andrews, Jr., interview, p. 12, 13.

[65]Alan Campbell interview with Bill Klein, p. 10.

[66]*Ibid.*

[67]*Ibid.,* p. 10.

[68]Ed Vandenberg interview with Bill Klein, p. 1.

[69]*Ibid.*, p. 2.

[70]*Ibid.*

[71]*Ibid.*, p. 6.

[72]English interview, p. 2.

[73]*Ibid.*, p. 3. At this time, this was exactly the same experience that Chuck Reed and other technical service people in the industry describe.

[74]*Ibid.*, p. 4.

[75]*Ibid.*, p. 5.

[76]*Ibid.*, p. 7.

[77]Conversation with Bob Finley, September, 1993.

[78]English interview, p. 15.

[79]Don Eyster interview with Bill Klein, p. 2.

[80]*Ibid.*, p. 1.

[81]*Ibid.*, p. 2.

[82]*Ibid.*

[83]*Ibid.*, p. 3.

[84]*Ibid.*, p. 5.

[85]*Ibid.*, p. 12. Dick Pearson says Avery "came up with the first computer label in the late 1950s, early 1960s." (Pearson interview, p. 10.)

[86]Jerry Nerad interview with Bill Klein, p. 1.

[87]Gray interview, p. 24.

[88]Nerad interview, p. 2.

[89]*Ibid.*, p. 3, 8.

[90]*Ibid.*

[91]*Ibid.*, p. 4.

[92]*Ibid.*, p. 6.

[93]*Ibid.*, p. 13.

[94]*Ibid.*, p. 14.

[95]*Ibid.*

[96]*Ibid.*, p. 8.

[97]*Ibid.*, p. 9.

[98]*Ibid.*, p. 10.

[99]Bud Gray Interview with Bill Klein, p. 1.

[100]*Ibid.*, p. 2.

[101]*Ibid.*, p. 3.

[102]Dan Tomlinson interview with Bill Klein, p. 9, 10.

[103]*Ibid.*, p. 3.

[104]*Ibid.*, p. 4.

[105]*Ibid.*, p. 8.

[106]*Ibid.*, p. 10.

[107]*Ibid.*, p. 12.

[108]*Ibid.*, p. 13, 14.

[109]*Ibid.*, p. 16.

[110]Andrews, Jr., interview, p. 21.

[111]Norman interview, p. 13, 14.

[112]Chuck Reed interview, p. 5.

[113]Bill Klein quoted in Bob Scheerer interview, p. 24.

[114]Bill Estridge interview with Bill Klein, p. 4.

[115]Jim Wilson interview, p. 20.

[116]Donovan interview, p. 8.

[117]*Ibid.*, p. 8, 9.

[118]Eagon interview, p. 7, 8.

[119]Gib Lewis interview with Bill Klein, p. 12.

[120]*Ibid.*, p. 15.

[121]*Ibid.*, p. 17.

[122]Campbell interview, p. 3.

[123]*Ibid.*

[124]Stillman interview, p. 12.

[125]Gerry Gartner interview with Bill Klein, p. 13.

[126]*Ibid.*, p. 13.

[127]Duane Hillmer interview with Bill Klein, p. 8.

[128]Stillman interview, p. 14.

[129]*Ibid.*, p. 15.

[130]Kyte interview, p. 25.

[131]*Ibid.*, p. 26.

[132]*Ibid.*

[133]*Ibid.*, p. 26, 27. Later in the interview Kyte says the "varnish" was really nitrocellulose.

[134]*Ibid.*, p. 30.

[135]McDaniel interview, p. 9.

[136]*Ibid.*

[137]Henry Anderson stated, "We're doing a lot of film printing these days. The pressure sensitive label business is not very large compared to what we're doing in other ways." Anderson interview, p. 8.

[138]McDaniel interview, p. 9.

[139]*Ibid.*, p. 9, 10.

[140]*Ibid.*, p. 10.

[141]Tom Wilson interview, p. 14, 15.

[142]McDaniel interview, p. 14.

[143]*Ibid.*, p. 12.

[144]First Fifty Years, p. 130.

[145]Paul interview, p. 23.

[146]Norman interview, p. 4.

[147]*Ibid.*, p. 4.

[148]*Ibid.*, p. 5.

[149]*Ibid.*, p. 7.

[150]*Ibid.*, p. 8.

[151]*Ibid.*, p. 15.

[152]Orlando interview, p. 12.

[153]*Ibid.*, p. 13.

[154]*Ibid.*, p. 15, 16.

[155]Richard Rosemann interview with Bill Klein, p. 5.

[156]*Ibid.*, p. 9. Tyvek and Mylar are registered trademarks of DuPont.

[157]*Ibid.*

[158]*Ibid.*, p. 11.

[159]*Ibid.*, p. 12.

[160]*Ibid.*, p. 13.

[161]*Ibid.*, p. 15

[162]*Ibid.*, p. 22.

[163]Eagon interview, p. 14.

[164]Maynard Louis and Les Ordman interviews with Bill Klein, p. 2.

[165]*Ibid.*, p. 6.

[166]Bruce Motter interview with Bill Klein, p. 17

[167]Louis and Ordman interview, p. 7.

[168]*Ibid.*, p. 8.

[169]*Ibid.*

[170]*Ibid.*, p. 11.

[171]*Ibid.*, p. 13.

[172]*Ibid.*, p. 19.

[173]*Ibid.*

[174]*Ibid.*, p. 20.

[175]*Ibid.*

[176]*Ibid.*, p. 24.

[177]*Ibid.*, p. 25.

[178]*Ibid.*, p. 22.

[179]*Ibid.*, p. 10.

[180]*Ibid.*, p. 10.

[181]TLMI Golden Anniversary publication, 1983, p. 3.

[182]Percival Wise interview with Bill Klein, p. 16.

[183]*Ibid.*, p. 17.

[184]TLMI, p. 3.

[185]Muir interview, p. 17.

[186]communication from TLMI, 1994.

[187]Herb Smith interview, p. 2.

[188]Beaudoin interview, p. 14.

[189]Klas interview, p. 18.

[190]Gene Singer interview with Bill Klein, p. 6.

[191]*Ibid.*, p. 7.

[192]*Ibid.*, p. 9.

[193]*Ibid.*

[194]*Ibid.*

[195]*Ibid.*, p. 10.

[196]*Ibid.*

[197]Klas interview, p. 17.

[198]*Ibid.*, p. 18.

[199]Buchta interview, p. 15, 16.

[200]*Ibid.*, p.17.

[201]George Parisi interview with Bill Klein, p. 7.

[202]*Ibid.*, p. 12.

[203]*Ibid.*

[204]*Ibid.*, p. 11.

[205]Long, Robert P., "Some of the Events Leading to FTA's Founding," FTA 25th anniversary article, p. 18.

[206]Parisi, George., "Flexography and FTA...'We've Only Just Begun'," FTA 25th anniversary article, p. 95.

[207]Andrews, Jr., interview, p. 32.

[208]*Ibid.*, p. 33.

[209]*Ibid.*

[210]Parisi, *op. cit.*, p 95, 96.

[211]Bob Scheerer interview with Bill Klein, p. 3.

[212]*Ibid.*

[213]Some interviewees recall a later date.

[214] Some interviewees recall an earlier date. Gummed rider labels had been around for decades by then, but successful use of PSARL required automatic application to be practical.

[215]Defined in the TLMI *Glossary* as "A thin, flexible blade mounted parallel to and adjustable against the surface [of] an engraved anilox roll for the purpose of scraping off excess ink or coatings."

[216]Scherer interview, p. 5, 10.

[217]*Ibid.*, p. 4.

SECTION FIVE
QUALITY CREEPS IN:
1970-1980

Previous comments notwithstanding, quality was not the critical factor in the 1960s. Availability was more important. Buy a new $5,000 press, or a $2,000 used one, and just try to keep up with demand. Although there were good salesmen in the industry, most would admit that salesmanship had less to do with getting the order than with being able to take the order and deliver it. The customers figured out what they needed and dangled money in the converter's face. "Make it for me—I'll pay." Most of what was requested was *do-able* and the converters and their suppliers enjoyed one record sales year after another. It was an entrepreneur's fantasy at its best.

The question of price is always relevant when new competitors enter a market. The new businesses were usually headed by the entrepreneur who was often the chief salesman and, in some cases, the printer as well. They were lean and hungry, had direct face-to-face contact with the customer, and provided excellent customer service. These advantages of the small entrepreneur quickly translated into rapid growth of the business.

In Europe and America double digit sales (and even profit) increases were expected and gained. Avery's Russ Smith spoke to the Los Angeles Society of Financial Analysts in 1965 when Avery sales were $30 million. Four years later he spoke again when sales topped $100 million. By the mid 1970s Avery's corporate sales were over $300 million, a tenfold increase in a decade. In fact, in 1974 the company, created atop a flower refrigerator thirty-nine years earlier, became a Fortune 500 company.[1]

In the same year a consulting firm hired by Avery said they saw no reason the twenty-five percent growth and fourteen percent return on equity enjoyed in the previous twenty years would not continue.

Base materials and release liners were—finally—beginning to achieve consistency. The web was wider, presses ran faster and were easier to run, the dies were more reliable. Automated labeling was practical, reliable, profitable, and the customers loved it.

There's an old salesman's joke that says quality, service and price—choose any two. Well, at the beginning of the 1970s the PSARL converters could offer service and price.

It was time to take a look at quality.

STANDARDIZE, STANDARDIZE, STANDARDIZE

In the early 1970s the Mark Andy Company, with Mark Andrews, Jr., now at the helm, made the decision to:

> *Build a standard machine that everyone can use. That was really my hope when I got in this business. My hope in the early 1970s was to discontinue building about thirty different models and come out with only four different models that would be standard models, and we would build those all the time and change them very infrequently....That's when we came out with the seven, ten and sixteen inch width.*[2]

Press speeds improved as well, going from 300 to 500 FPM by 1976. Andrews doesn't see much competition in the United States from off-shore companies even today. There are some—Gallus, Ko-Pack and Nilpeter, but they compete much more strongly in their own part of the world. There are some differences:

> *For instance, one of the things Europe is very strong in is flat bed die cutting. The Orient, too—Japan is very big in flat bed die cutting. Here...everything is rotary.*[3]

The reason? Unlike North America where rotary die manufacturers are in every production center, such had not been the case overseas.

In 1972, Mark Andy introduced an electronic feedback tension system:

> *It's not so critical on central impression presses...because the web is more or less clamped to the impression cylinder while it goes through three or five colors. But when you start building in-line presses and the web stretches out for a hundred feet...we're talking to the thousandths of an inch. We found the only way we could do that was with a very accurate tension control.*[4]

THE SWISS CONNECTION

The Gallus™ presses have offered some significant European competition to the American manufacturers since Gallus introduced its central impression flexo press in Europe in 1969, and by 1974 marketed the first high speed rotary letterpress.[5] Ferdinand Rüesch, Jr., of Gallus, notes that this first rotary four color letterpress was sold to Dennison Manufacturing. "We have been able to integrate the so-called flex die into a rotary machine. That was our innovation which made Gallus really grow." [6] Although expensive, the quality of the output is exceptional. However, by 1975, four color process flexo printing began to come of age and primary labels became fair game for flexographic printing.

> *[From] '68 to '72...we made the fundamental changes from the pop and mom shop to a highly industrialized technology driven company....From the slow running machine...basically the rubber flexo...a high degree of sophistication came into our presses.*[7]

Gallus continued to improve, and the first company in the United States which moved to the new *eight* color rotary letterpress was Soabar, part of Avery.

> *There was a funny incident when they [Soabar] came over [to Switzerland] to accept the press. They shipped the material over for the test stock...and we experienced registrations problems. So we said, "Oh, God, what do we do?" [We had a brainstorming session and] people said, "You know we have to make sure that the theory of climatization will be the same." With that we took it and went into a [controlled] climate to make sure...came back...put it on, and my God, it was running perfectly.*[8]

By now Gallus had become the main supplier for the letterpress technology world wide, as well as Europe's flexographic leader. The letterpress, though, was three to three and half times more expensive than a flexo machine of the same width and construction. Then, however, the difference in quality was great. "You could not produce [on] a flexo, just simply not what you could produce on letterpress." [9] Rüesch credits, among other things, the consistency of the letterpress.

Often, early introduction of new technology can face customer resistance. In 1979 when the first ten color fully rotary letterpress was introduced, even Gallus had some problems. According to Beda Künzle (who spent thirty years with the company and is responsible for the "Gallus History"), at the first showing of the R 160 B, one customer judged the machine to be unsellable. Why? Because it was "over-engineered!" Beda further points out that that particular customer is now the owner of twenty R 160 machines, and that the R 300, introduced in 1990, not only includes ten colors, but is a combination processing unit: letterpress, flexo screen, hot foil and varnishing.[10]

FROM SUB TO INDEPENDENT

Mark Herrman started building presses as a subcontractor for Jim Brown's International Machine Products. "Around 1974 a company by the name of Northern Engraving Company purchased National Tape and Label and all its subsidiaries" [11] (including International Machine Products). Northern Engraving was a metal name plate company and was losing ground to pressure sensitive and foils, so it bought Brown's pressure sensitive businesses in order to generalize. Herrman decided that this was an ideal time to strike out on his own.

> *Mark Andy was controlled by Mark Andrews, Sr., at the time [and built] all kinds of things and no two were alike. That was the reputation. The Webtron was the machine that everyone placed on a pedestal as far as flexo went....But we did things like 360 degree registration...a lot of special challenges....That led us into the road of special products and that's why we didn't become a larger company some years ago.*[12]
>
> *We tried to improve the quality of the machine, that was our goal, to put higher finishes, higher concentricity tolerances, better design. I think where the evolution in quality began was really toward the later 70s....Plants didn't have climate control back then....We didn't have the sophistication in the systems then.*[13]
>
> *We couldn't build machines and stockpile them, because everything was different. So my idea originally was [to] build a modular machine and just build modules, a giant erector set. Let's build unwinds, let's build printing modules, die cutting stations. Then whatever the customers desire we could provide.*[14]

So in 1975 Herrman's Custom Machine Company became Comco (CONverting Machinery COmpany). Instead of special projects and custom engineering, Comco began to focus on the pressure sensitive industry. And, as the label industry grew in the 1970s and 1980s, Comco grew with it.

By the mid 1980s Comco was searching for a letterpress partner in order to broaden the range of the business. At the same time, Gallus was looking for a flexographic partner. The two joined forces in 1987, and until 1992 Gallus was the world-wide distributor for Comco. Now, Herrman says:

R. Stanton Avery

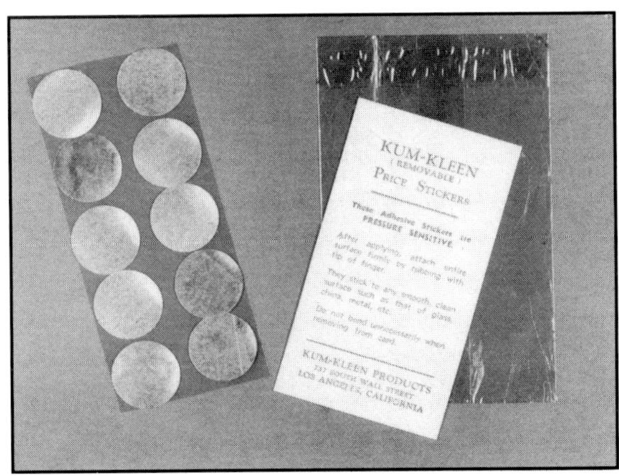

First self-adhesive diecut labels, 1935.

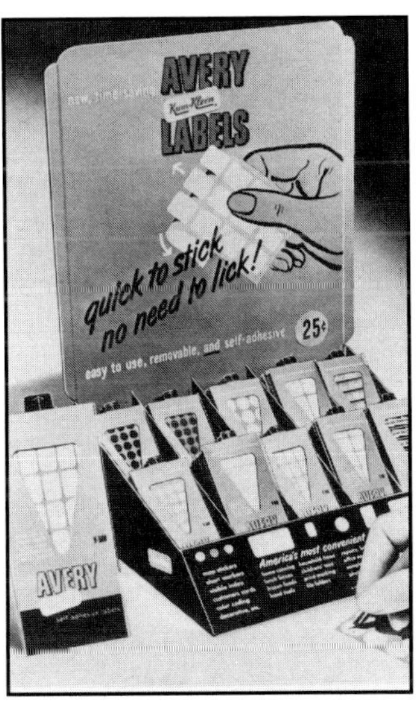

Kum-Kleen label display, 1945.

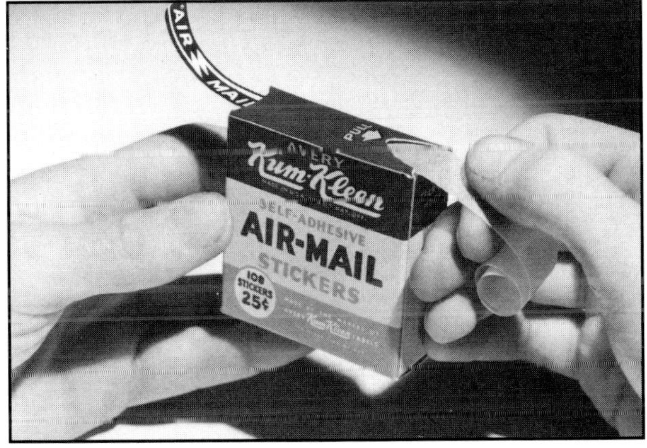

Well known psa label dispenser box.

All photos courtesy of Avery Dennison.

Mark Andrews Sr.

Early Mark' Andy self wound tape printing press.

Below, Mark' Andy roll label presses of the late '50s & early '60s.

All photos courtesy of Mark' Andy, Inc.

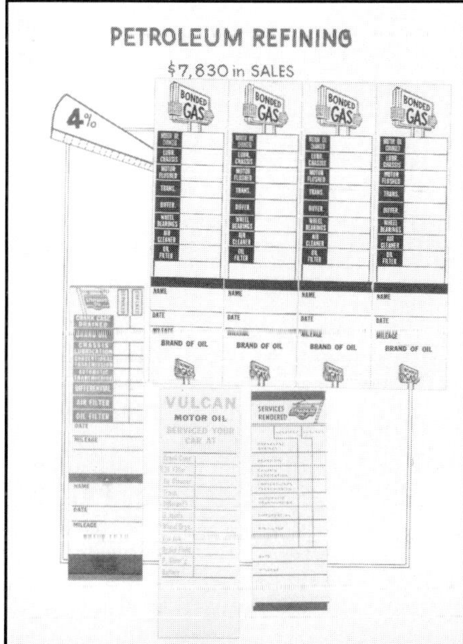

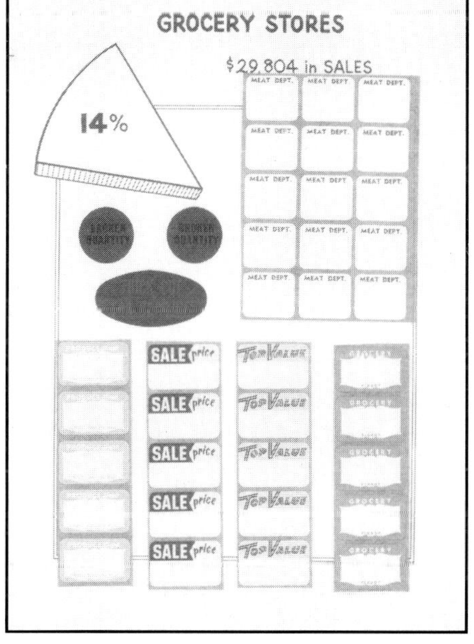

Pages from a Dennison salesman's sales manual illustrating Pressure Sensitive adhesive roll label applications of the late '50s and early '60s. *Courtesy of Walt Webster.*

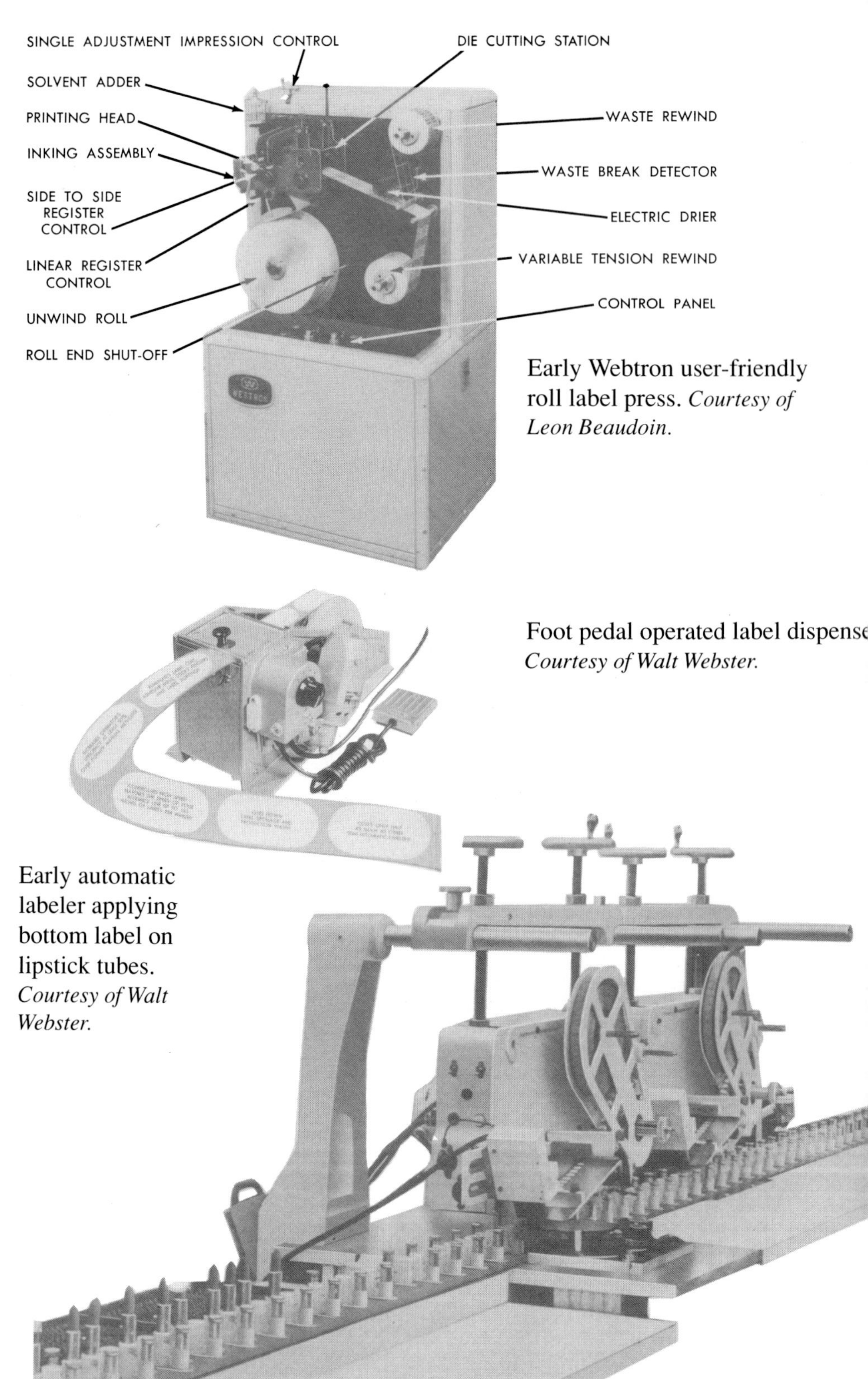

Early Webtron user-friendly roll label press. *Courtesy of Leon Beaudoin.*

Foot pedal operated label dispense
Courtesy of Walt Webster.

Early automatic labeler applying bottom label on lipstick tubes. *Courtesy of Walt Webster.*

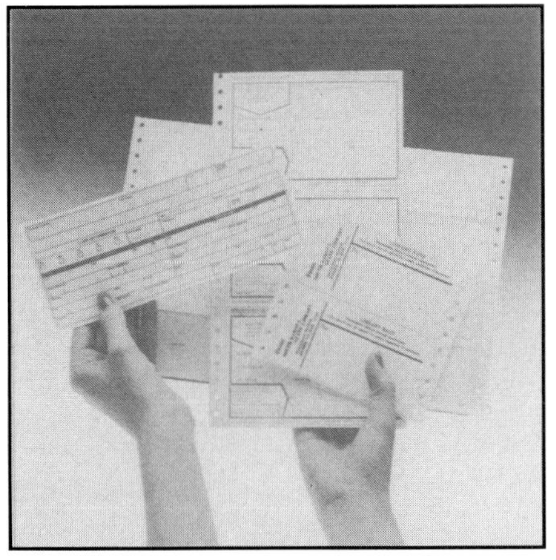

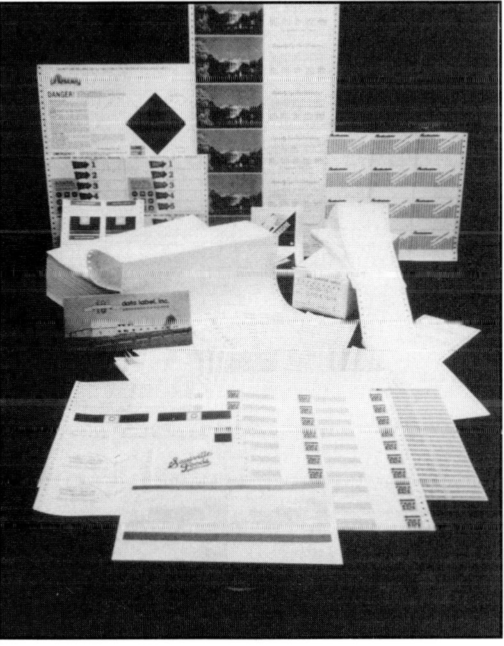

Pressure sensitive adhesive Electronic Data Processing (EDP) labels for a wide variety of applications, including the original application as address labels. *Top three photos courtesy of Avery Dennison. Photo on left courtesy of Data Label.*

Brand indentification labeling. *Photo courtesy of Bill Klein.*

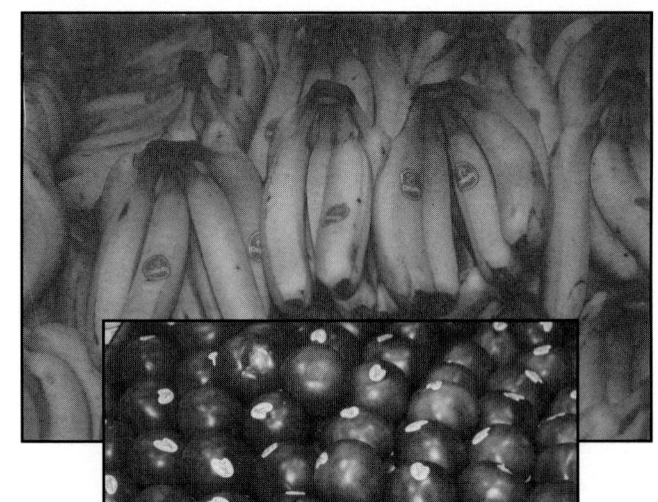

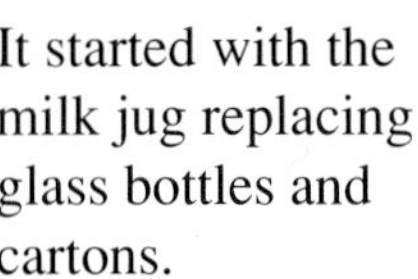

It started with the milk jug replacing glass bottles and cartons.

The famous sock label application.

Photos courtesy of Bill Klein.

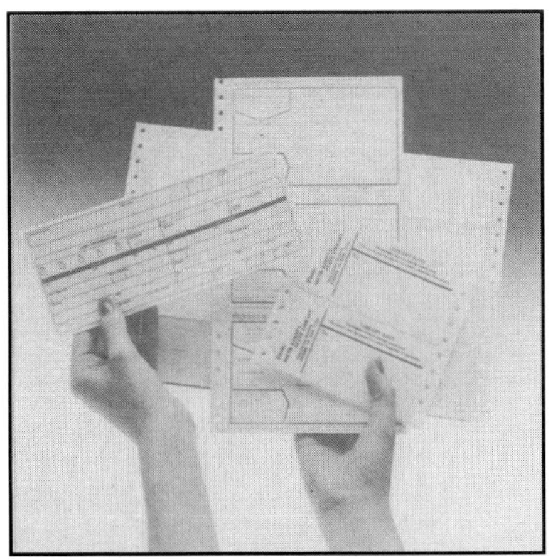

Pressure sensitive adhesive Electronic Data Processing (EDP) labels for a wide variety of applications, including the original application as address labels. *Top three photos courtesy of Avery Dennison. Photo on left courtesy of Data Label.*

Brand indentification labeling. *Photo courtesy of Bill Klein.*

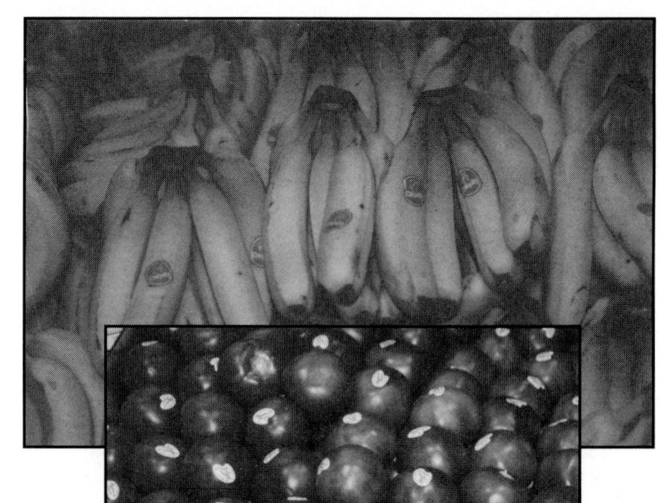

It started with the milk jug replacing glass bottles and cartons.

The famous sock label application.

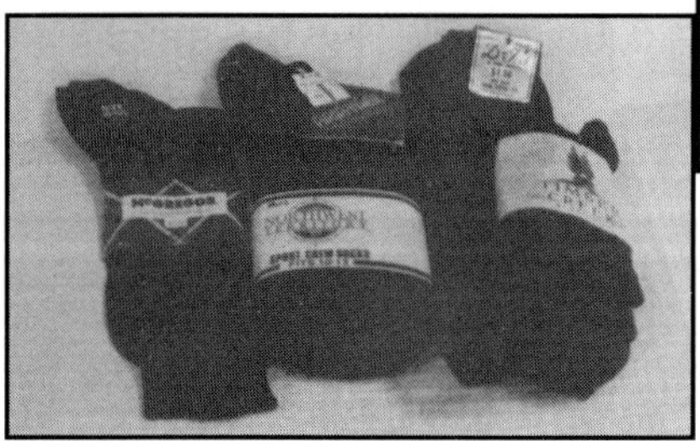

Photos courtesy of Bill Klein.

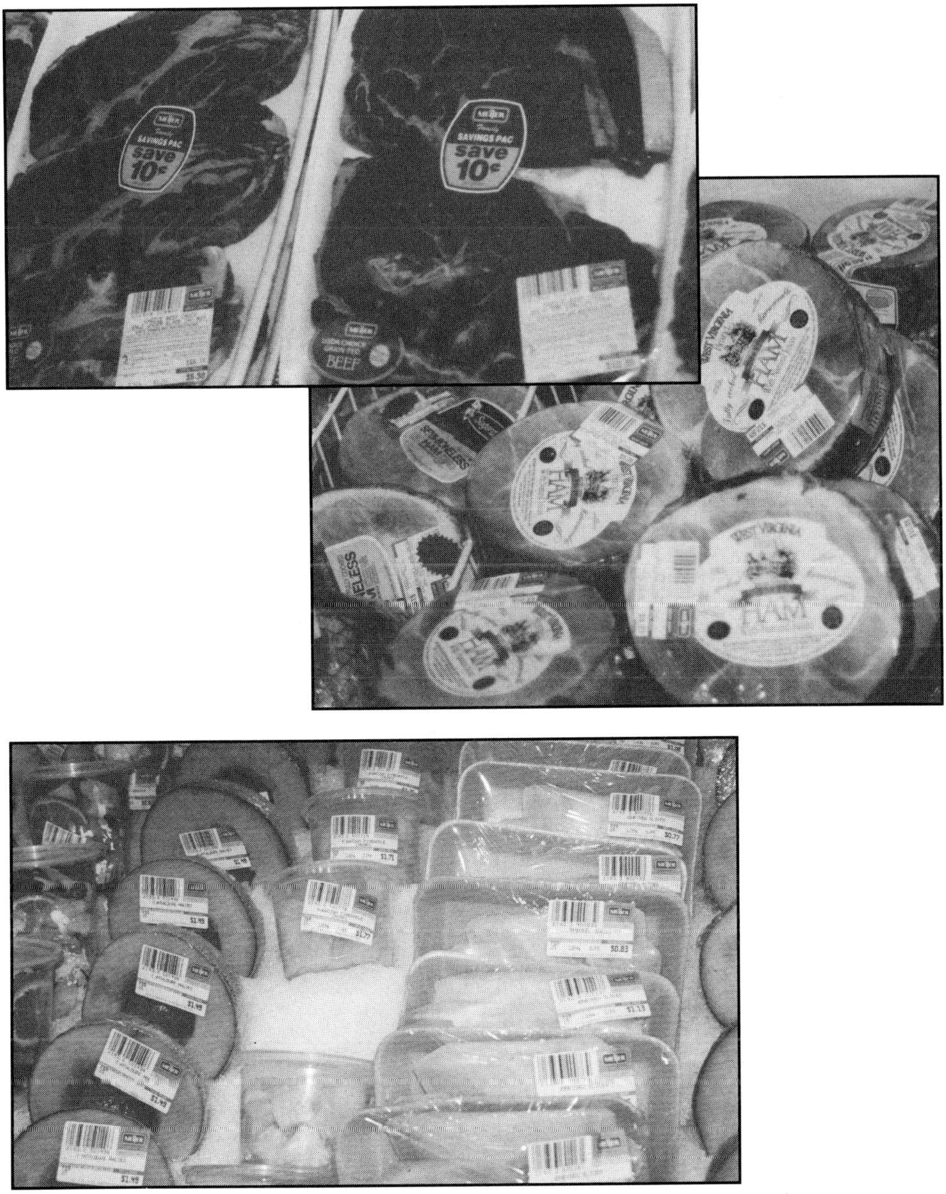

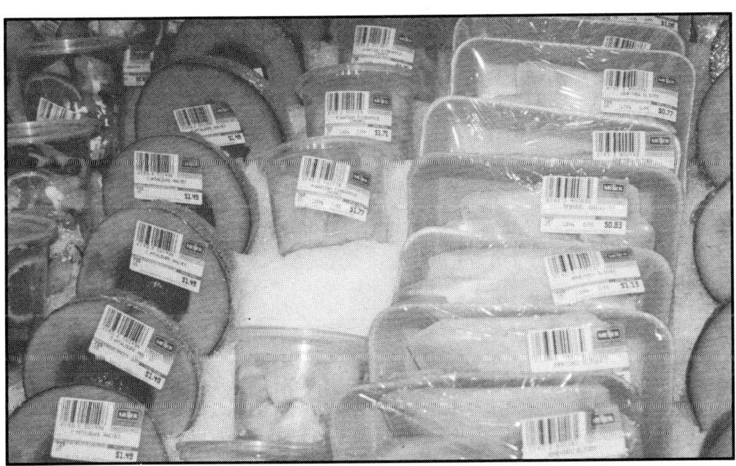

PSA labeling in supermarkets is extensive - brand identification, variable weight, UPC and special price labels shown above. *Photos courtesy of Bill Klein.*

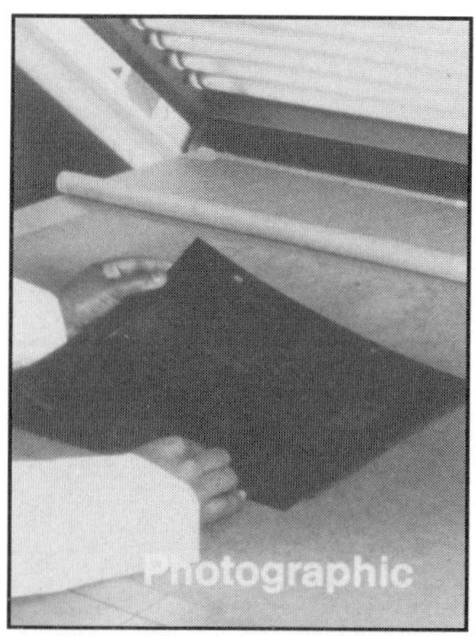

Processing photopolymer plates. *Photo courtesy of E. I. DuPont.*

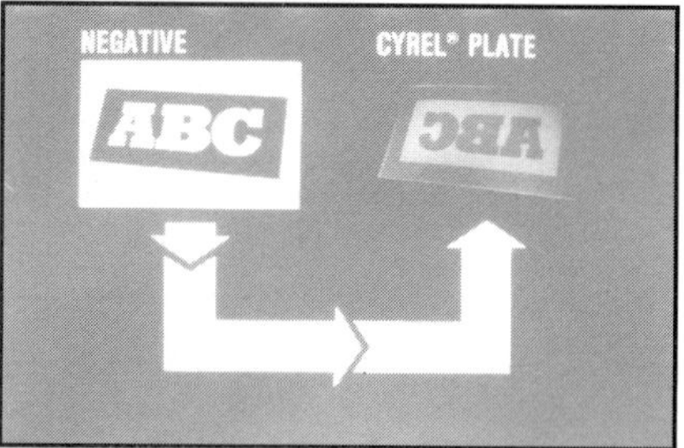

Printing with polymeric plates and rotary die cutting and stripping of matrix. *Photos courtesy of Gardoc.*

Mounted polymeric plate - anilox roll and doctor blade. *Photo courtesy of Gardoc.*

Hotpoint
Hotpoint
Hotpoint

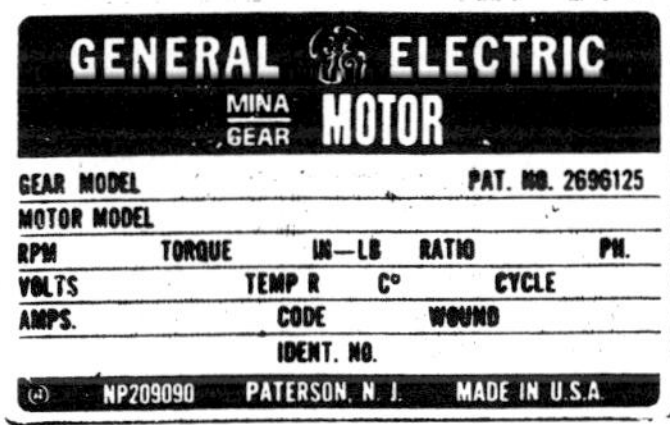

Pressure sensitive adhesive labels from the 1960s - used for brand identification. *Samples courtesy of Bill Klein.*

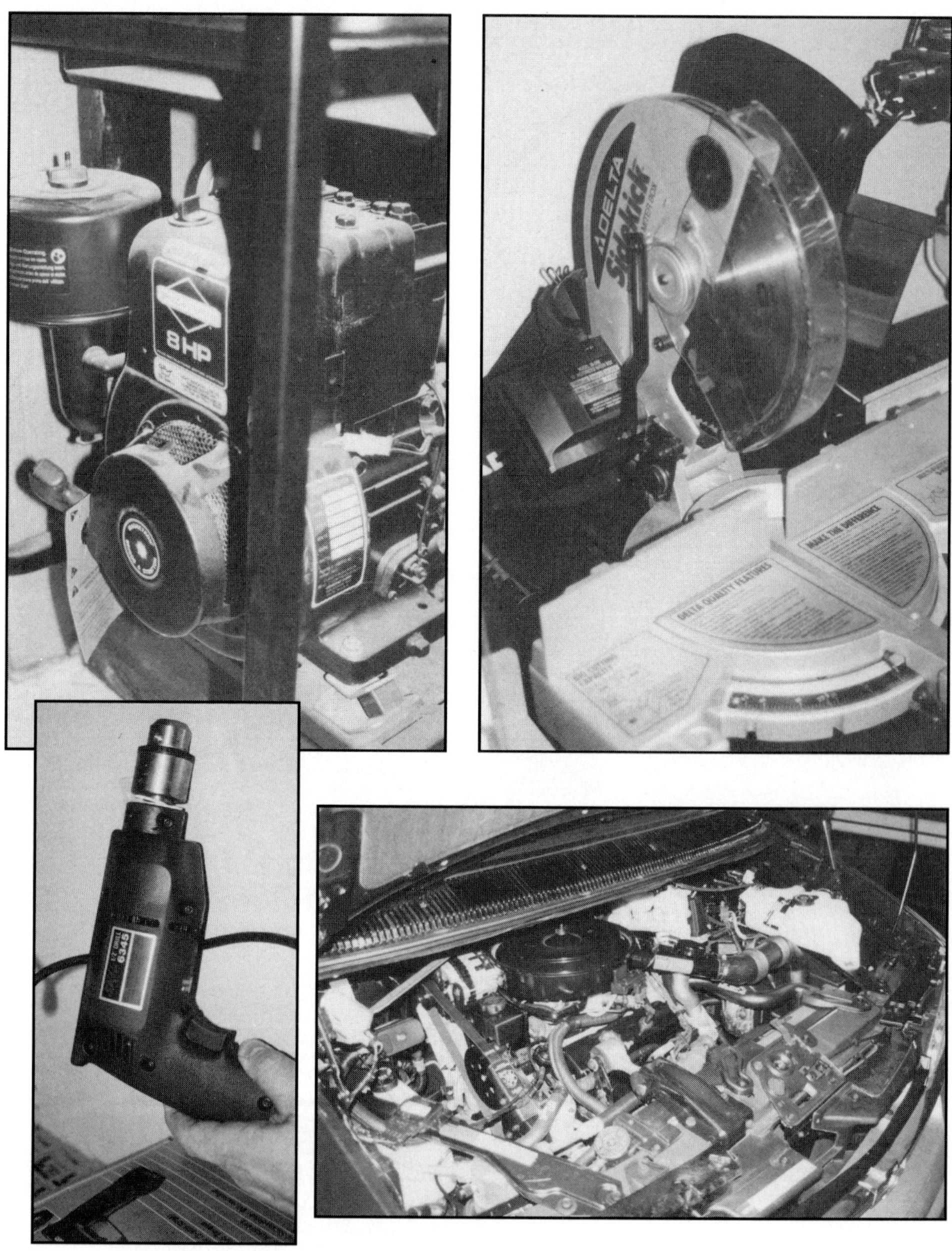

Industrial label applications continue to expand - there are 32 separate psa labels under the hood of this Chevrolet van. *Photos courtesy of Bill Klein.*

Photo courtesy of MPI.

Dramatic illustrations of the broad range of food and other consumer products identified by pressure sensitive roll labels.

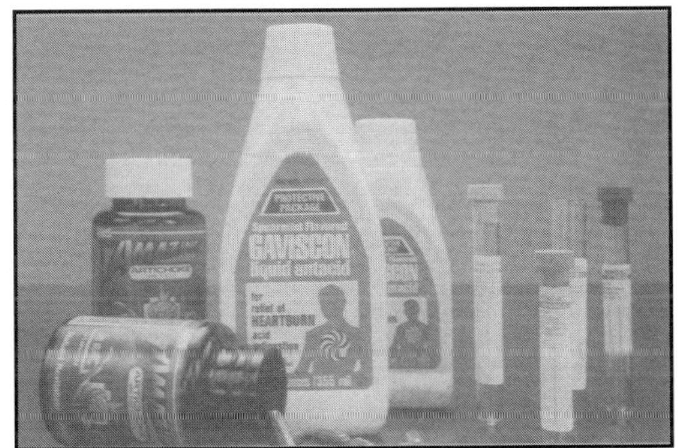

Photo courtesy of Package Service.

Photo courtesy of Lord Label.

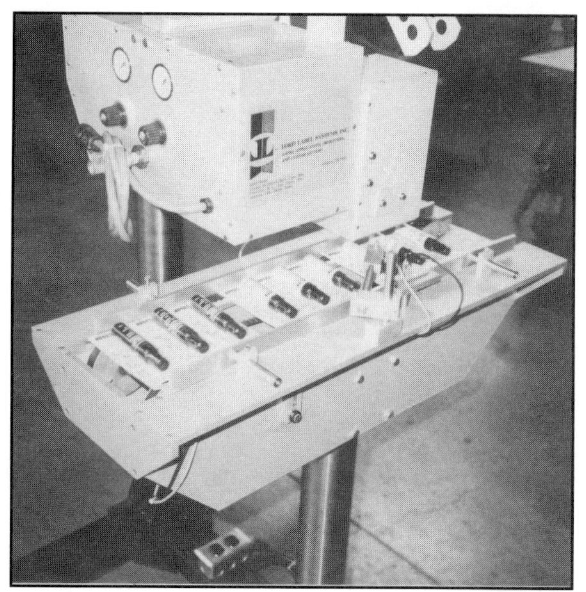

Modern automatic systems for applying pressure sensitive adhesive roll labels. *Left photo courtesy of Lord Label. Bottom photos courtesy of MPI.*

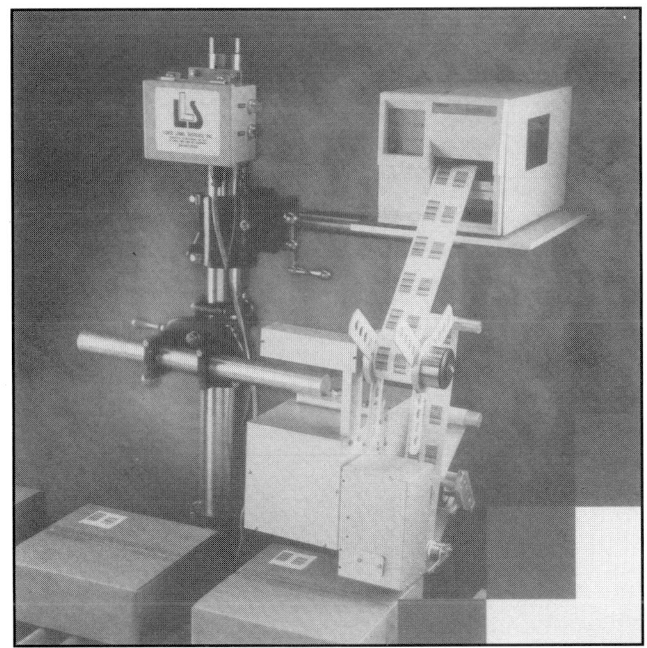

Bar Code Printing and automatic psa labeling system for commercial/ industrial applications. *Photo courtesy of Lord Label.*

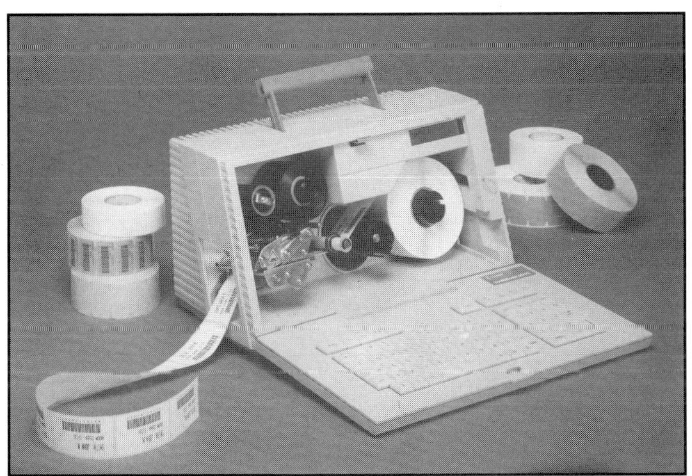

On demand printing of psa bar code labels of medical pharmaceutical applications. *Photo courtesy of Timemed.*

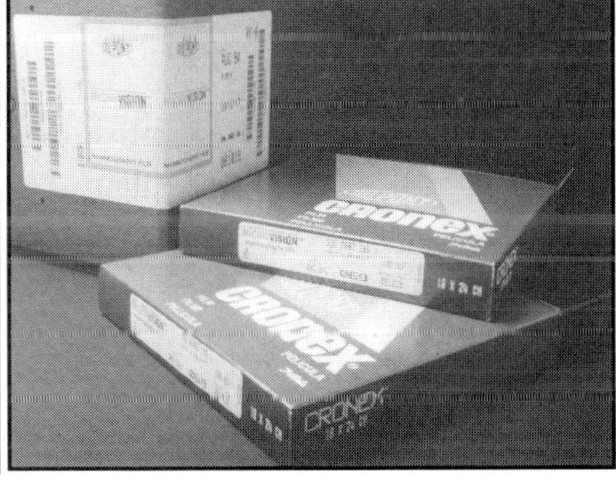

Commercial bar code applications. *Photos courtesy of Avery Dennison.*

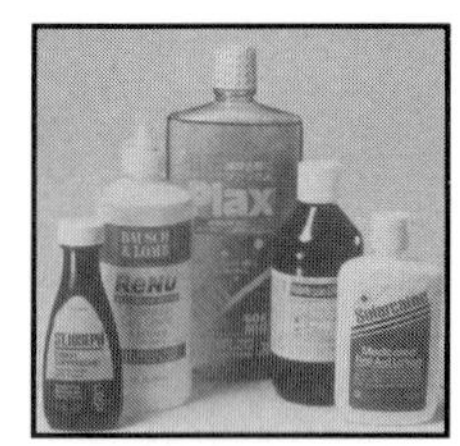

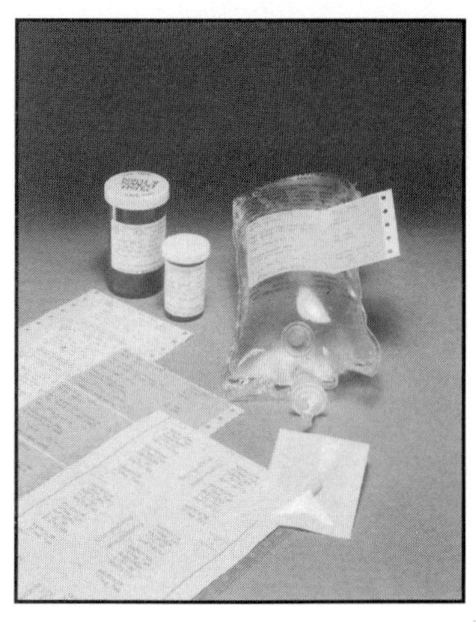

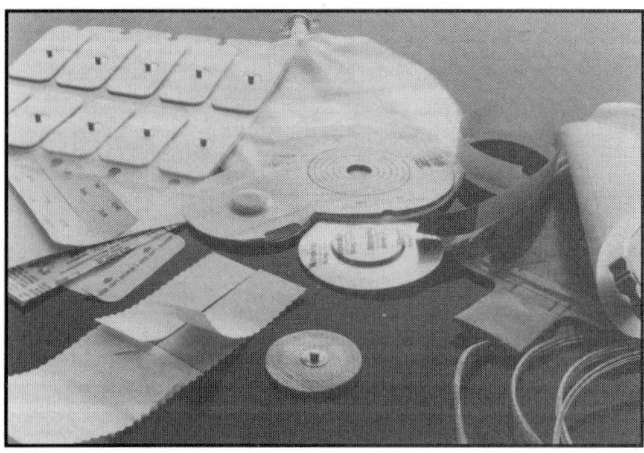

Pharmaceutical and personal care applications of pressure sensitive adhesive materials. *Photos courtesy of Spear, Engraph, Timemed and Avery Dennison.*

Comco Seriflex

Mark' Andy 2200

Webtron 1000

Gallus R200

Modern Presses for the production of psa roll labels.

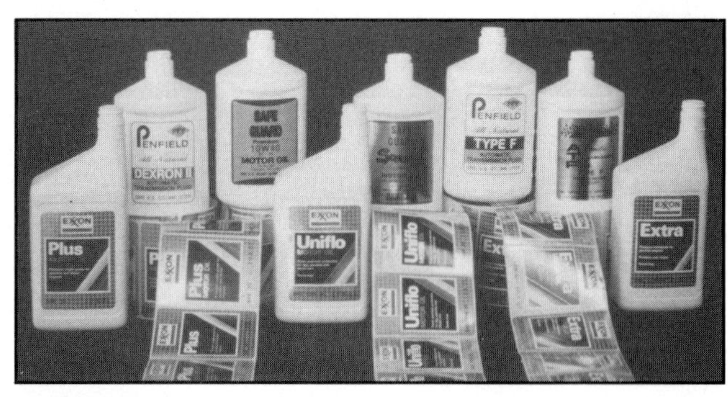

Photo courtesy of Gardoc.

Photo courtesy of Spear.

Photo courtesy of Avery Dennison.

Photo courtesy of Engraph.

Decorative and prime labels on plastic containers.

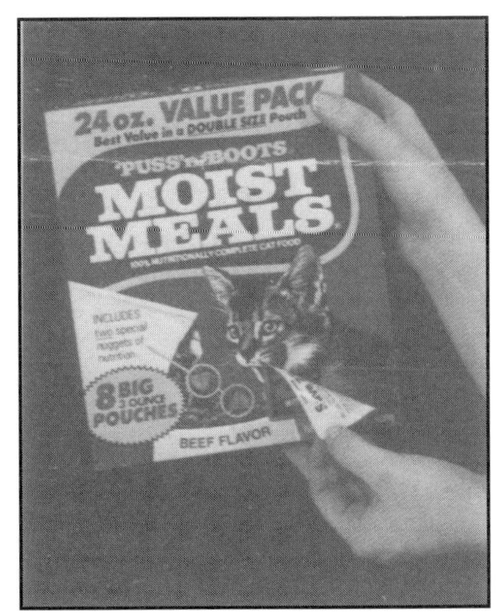

Top photos courtesy of MPI.

Photos courtesy of Engraph.

Inserts, Instant Redeemable Coupons and Premium PSA Label Applications.

Requirements for nutritional information required doubling the size of this milk jug label.
Photo courtesy of Bill Klein.

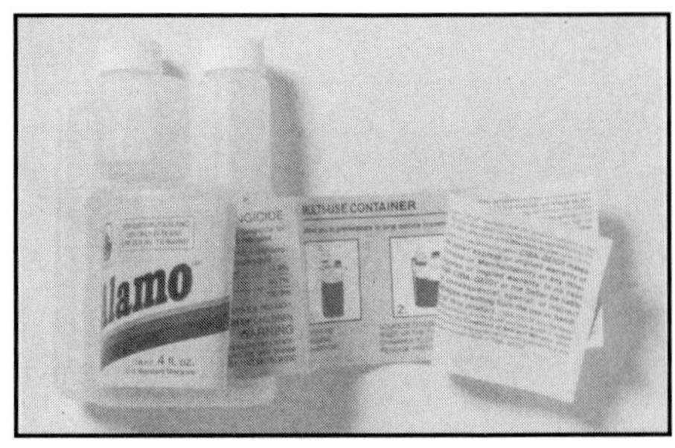

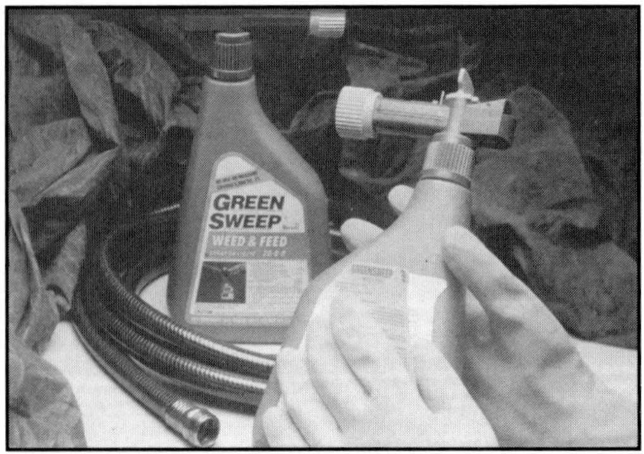

Folded Multi-panel (extended information) labels.
Top photos courtesy of Engraph.
Lower left photo courtesy of CCL.

The "No Label Look" - *upper photos courtesy of Engraph. Lower photos courtesy of Spear.*

Security National Bank
40 South Limestone Street
Springfield, Ohio 45502

WILLIAM KLEIN JR
6200 PLATEAU DRIVE
SPRINGFIELD, OHIO 45502-9233

PRIN116 927051014 1N 02/12/94
RETURN TO SENDER
NO FORWARD ORDER ON FILE
UNABLE TO FORWARD
RETURN TO SENDER

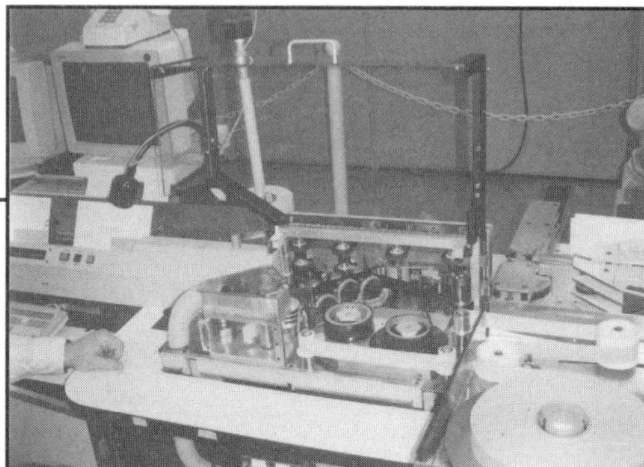

Linerless psa label applicating system for US Postal Service CFS 11 labels - *Photo courtesy of Cordant.*

Pressure sensitive postage stamps. *Photo courtesy of Avery Dennison.*

PSA hologram security label. *Photo courtesy of MPI.*

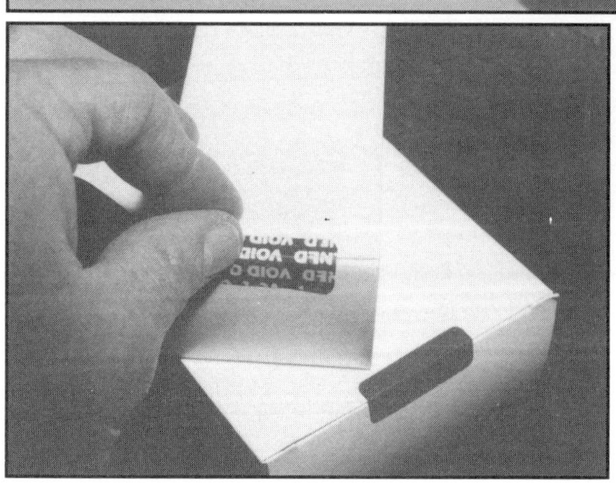

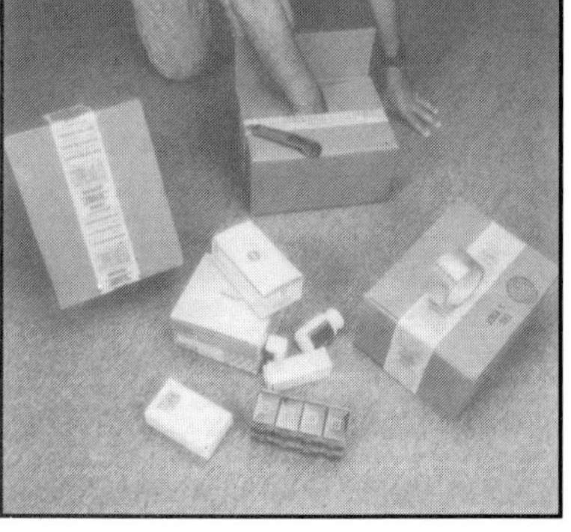

Pilfer indicating labels and tapes. *Photos courtesy of Graphic Materials Inc.*

In 1992 we decided mutually to go back on our own....I think the real opportunity today is to provide something the competition can't....We're right on the verge...of electronic printing, plateless printing, let's call it. Direct from the computer to some kind of an imaging system.[15]

One of the more recent developments by Comco is a chill drum for their printing modules, to be used with films and UV inks.

CARLOADS OF TWO MIL CLEAR

Only a few companies will hire someone at 52 years of age. Mark Ungerer of Flexcon hired Walt Webster who had been twenty years with Dennison.

One of Webster's earliest success stories came as the result of tragedy, when a killer introduced poison into bottles of the painkiller Tylenol™. "We sold literally carloads of two mil clear acetate."[16] The acetate created a product that would clearly reveal if any tampering took place between the production line and the customer.

Webster recalled a story from his earlier days with Dennison, perhaps about 1960, when he made a sales call with the top tag salesman in Cincinnati, Ohio.

I don't think tag salesmen die, they fade away. I remember getting off the plane and George approached me. A hell of a nice guy. He said, "Gee, Walter, sure glad to make your acquaintance...but we don't have any label business out here." That's exactly what he said. Now he and I make calls on GE and...this guy at purchasing had offered me a chance to make the trip around the plant. [The Purchasing Agent asked], "Do you have any ideas?" I said, "You're hand-writing all these labels out (they were doughnut shaped pressure sensitive). Do you realize that these could be die cut with a liner and they could be run through your computers when you place your orders for materials?" George says, "I tell you...we can make a five part tag and we can have a string on each one and we can tie them on." I almost fell under the table....It shows you how deeply these things are in everybody's mind, these stinking tags. The young people who came on board, no problem. But the older guys, I hate to say it being older myself, found it more difficult to switch to that label line.[17]

When Webster joined Flexcon, sales were about two million dollars. When he left in 1984 his sales alone were more than ten million. When he left he was controlling about ten percent of Flexcon's business and "they were a little bit concerned about me going somewhere else. I said, "You hired me at 52, I ain't going to do nothing against you at all."[18]

SHOVED INTO THE MARKETPLACE

George Noah of Lewis Label Products was silk screening pressure sensitive bumper stickers when he was eleven years old! It was in his father's part-time business. The senior Noah was a fireman, 72 hours on, 72 hours off, and the business kept both busy until it was sold after Noah graduated from high school.

A paper merchant salesman, Gib Lewis had realized that there were no small label companies in the Dallas/Fort Worth area, so he founded one in 1964, and Noah began working for him in 1965. After five or six months, Noah left and spent three years elsewhere, then returned to Lewis Label Products in 1968.

> *I would sell during the day and then print at night. Gib would help me with the screen printing....Price marking labels mostly...not into prime labels at all. I think in 1978 a customer approached us about some color process work. We attempted it on flexo and you just couldn't get it flexo.*[19]

In 1979 Lewis Label Products bought a Ko-Pack rotary letterpress. "So," Noah says, "that's what kind of boosted us into color process prime label business in 1979. And it's just kind of taken off ever since."[20]

Noah says he has really had more supplier push than market pull.

> *Because people like Fasson...MACtac...were pushing their products, 3M was going to the consumer, saying, "Here's my material, here's what it can do," and say, "Lewis Label Products can do this." My customer [would come to me and say], "3M said you could do this." And... they almost shoved us into the marketplace.*[21]

Noah has been with Lewis Label Products for more than 27 years. Lewis Label Products has four employees with more than 27 years and twenty employees with more than fifteen years service. Gib Lewis himself is not around on a day-to-day basis, however. He's a member of the House of Representatives of the Texas legislature in Austin, and has been Speaker for the last ten years.

UNDER THE HOOD

Bob Smith has worked all around the United States. He had been working for Paramount Paper Products while going to the University of Nebraska in Omaha.

When I graduated I thought there was a lot of opportunity in the pressure sensitive business in 1961....I was moved to the east coast...to run their eastern operations. They were acquired by National Corporation and I worked within National for a short time...then moved to a new job [with] Alan Hollaender. And they [opened a midwestern plant in] Chicago. I ran the midwestern operation for them... and Litton Industries...acquired Alan Hollaender.[22]

Litton had acquired the Kimball Tag business at about the same time they bought the Hollaender operation and had Smith run the Kimball Tag business for them. Smith recalled 1967 or 1968 as the date. But then he was recruited by Avery and moved to California. His moving days were over as he has been with Avery/Dennison ever since. His first job was manager of Avery Label's own base materials manufacturing operation.

I worked on the materials side for a few years...and then became director of operations with responsibilities for all of their manufacturing plants here in the United States and their engineering group. [23]

Smith said the biggest advantage of pressure sensitive is that it will work with such a wide variety of substrates. He adds:

One thing that Avery did and has really enjoyed the benefit over a long period of time was their pioneering effort in the materials that could be used...[including those for] extreme temperatures and some extreme conditions. Like the automotive industry and the under-the-hood applications. If you buy a new car today, if you raise the hood, it's amazing the pressure sensitive labels that are there.[24]

Some cars reportedly carry more than a hundred pressure sensitive labels. Some are safety instructions, others are for identification or tracking purposes.

Smith pinpoints an area that will continue to create more growth in the industry:

> *That is the desire of the marketers to differentiate their product more and more, and to have more eye-catching appeal on the product. For example, the beverage industry bottle—if you look closely—pretty straight forward [gum label] graphics today. With the opportunities for pressure sensitive and heat transfer products we can decorate with multiple colors, far more eye-catching appeal on these....Obviously this will begin in the higher-priced limited quantities brands. But it could grow to be more in the mainstream and the volume there is just tremendous.*[25]

MORE BASE MATERIAL IMPROVEMENTS

The base material suppliers were also doing their part to come up with more new and better quality pressure sensitive adhesive coated materials to meet the growing needs of the roll label printer. Release liner technology continued to improve, the adhesive coaters were adopting better control systems, and the paper and film manufacturers were beginning to improve the quality and consistency of the materials sold to the adhesive coaters. The result was a growing variety of materials to meet the growing requirements for specific label applications, including low and high temperature adhesives and unique substrates.

In 1973 Fasson's AT1 established a pattern of use, both as a general purpose permanent adhesive and as a cold temperature adhesive. In 1975 a new adhesive polymer was introduced by the Shell Chemical Company. Kraton was a thermalplastic that could be applied at high speeds using the hot melt technique. In 1975 Dow Corning introduced solventless silicone technology, providing greater consistency of release and lower release levels. Each of these introductions helped improve the quality of the product.[26]

AL NORMAN ON THE TECHNICAL BREAKTHROUGHS IN THE 1970s

From Al Norman's interview with Bill Klein:

I think another breakthrough in the industry was the introduction by Shell Chemical of Kraton™ polymers in the mid 1970s. That launched hot melt....It was EPA compliant...because there was no solvent involved.In the mid seventies Dow Corning [introduced] a solventless silicone which gave a release system that could be cured at high speeds because you didn't have to dry any solvent off. That was very compatible with hot melt technology....The big difference was if you want to run a solution coater at the speeds you run hot melt coaters, you'd need something that looks like an aircraft carrier, because you've got to have those very long drying tunnels....With hot melt, coating speed is simply a matter of mechanical design.The kind of polymer development and the quality of the polymers that we really know today also started to emerge around the mid seventies. And...it was driven by the EPA....Otherwise we'd still be coating with solvents.A lot of people don't understand this but all of the major guys [base stock manufacturers] put in solvent recovery, it was not just for EPA compliance, it was that we didn't have to confront our customers and tell them they had to start all over again with totally different raw materials with different problems. So I've always thought solvent recovery was more a marketing protection strategy than EPA compliance, even though you got the benefit of that [reduced pollution].

A DIFFERENT MIND SET

In 1970 or 1971 Bev Eagon accepted an offer from MACtac and returned to northern Ohio. While Fitchburg had tried to duplicate Fasson's product line, Eagon said MACtac did not.

But some things he [Morgan] did better than Fasson... he saw things Fasson didn't see. For example, he was heavy into industrial and Fasson wasn't even in the ball game. One time he said he'd do the Chinese laundry if there was any money in it. Burt could take [industrial tapes] and make a silk purse out of a sow's ear.[27]

While at MACtac, Eagon worked on reflective materials (road signs, bicycle markings, etc.), heat seal adhesives and fire retardents, "almost all solvent-based adhesives." [28]

WUTHERING HEIGHTS

In 1976 Eagon left MACtac and spent a year caring for her dying mother. Then she joined Brown Bridge, a Kimberly Clark company.

She had turned them down once, partly because when she visited the plant, she said, "Man, this plant is something out of *Wuthering Heights*. It was the smartest thing I ever did. I thought Kimberly Clark was the nicest company I ever worked for." [29]

Although she liked the people, she felt the company was never as knowledgeable as it should have been. When she arrived they were not coating their own silicone paper. "They didn't even know they should....The day I calculated out [the savings] when they saw those figures they almost went bananas."[30]

Finally Kimberly Clark approved the expenditure and the equipment was purchased. Then, says Eagon, "We had the fastest start-up time ever for silicone release paper." [31]

Brown Bridge had a meeting every day, says Eagon. I said, "This is like an AA meeting. You all come in and support each other and you decide to come back the next day." [32]

Would she do it all over again?

I think so. Other people have asked me that. I've always said the biggest mistake I ever made was leaving Dow Chemical. And, in many ways, that's right. But on the other, I wouldn't have had such an interesting career....You don't get many opportunities to be in on the birth of an industry....There was something else about this industry, you got to know everyone. After a while, even if they were only at meetings, once a year you got to see them. And then you talked with them on the phone. I have friends who will call me that I haven't seen in five or six years. And my phone will ring and [someone will say], "Eagon, how you doing?" We talk just like we saw each other last week.[33]

FROM PUSH TO PULL

Bill Estridge, who had joined Fasson in 1961 as a sales representative in the southeastern states, became Fasson sales manager in about 1970, and initiated a change in Fasson's sales philosophy.

In the 1960s they had been selling label materials to the converters. In the 1970s they created a program called "Specification Selling," and began calling on the end users and helping them create *specs* for labels. Such specification selling, as you might imagine, favored pressure sensitive in general and Fasson pressure sensitive materials in particular.

And also I was pushing the total systems approach. What was happening to a lot of customers was that you just didn't have a total selling situation. The converter could go in and talk to them about labels, but he didn't know a whole lot about adhesives. So sometimes we would go in with the converters, talk to them about adhesives, what the product would do, what they wanted it to do, and then the next question was, "Well, how are you going to get the labels on?" If we couldn't answer that question then we were probably going to lose the sale. So with the Fasson Machine Group, then we had to say, "We can handle that too."[34]

Of course, this was an industry-wide program, requiring a lot of time out in the field. Estridge remembered traveling to the west coast as often as three times a week. "There had to be a good reason. The food was good."[35]

BETTER THINGS FOR BETTER PRINTING THROUGH CHEMISTRY

Almost without exception the one hundred plus interviewees who *wrote* this book cite four major contributors to the industry—Avery, Andrews, Mages and DuPont.

Although the glass reinforced plate was introduced in the mid 1960s, DuPont had introduced the first polymeric plates in about 1950 for letterpress and offset presses. It wasn't until 1972 that they introduced FR, Flexographic Red. The narrow web flexo printer could now produce process-color labels of greatly improved quality. DuPont marketed FR photopolymer plates to both converters and press manufacturers.

Peter Menzian and Tricia Basto of DuPont were jointly interviewed:

> *BASTO: We could provide a more uniform plate than they were ever able to get with rubber.... They were getting used to doing what was unheard of, process work....The presses were getting better at the same time. You had improvements in the plates and presses right at the same time.*[36]

When GarDoc was getting started about the same time as the FR photopolymer plate was being introduced, Menzian says:

> *GarDoc wouldn't have started if it wasn't for some of the suppliers like ink companies, plate companies and press companies to say, "Yes, this industry is going to grow, we'll help you financially by pledging equipment, and we'll give you lots of time to repay.*[37]

How an ink company, a plate company and a press company helped launch GarDoc is covered later in this section.

DuPont enjoys a reputation for continuing to improve quality. The FR photopolymer gave way to the HL and there have been specific new products to meet changing needs. DuPont's current products include Cyrel™, which meets the international ISO 9000 certification.

Tricia Basto was working on UPC codes in 1972 and there was some doubt that flexo could produce readable code. The photopolymer plate was the answer, and would drive the sales of the revolutionary product leading directly to primary label business.

DuPont, naturally, wanted everyone to know about their printing plates, so they emulated Vandenberg by holding seminars:

BASTO: We had every month training classes [for] plate makers to the supervisors. [We taught] beginning and advanced [classes. These classes] were full every time, tag and label [producers] probably making up about sixty to seventy percent of our attendance.

MANZIAN: Education played a key part in promoting flexo quality.[38]

OBSOLETE ON ARRIVAL

Although the glass modified plates were superior to rubber, the plate was still the weakest link in the flexographic process. Mark Andrews, Sr., decided to eliminate that weak link by developing a machine that was both a true rotary letterpress and a UV cure litho press, first called the G Press, then the 5100. John Garber and his father, John Ruskin Garber, worked with Andrews on the machine.

Ultraviolet curing had been used by a few newspapers. In 1972 there was only a single source for UV lamps and a single source for UV inks. Then photopolymer plate conquered the weakest link in the flexographic process and quality flexographic printing was within reach, making Andrews' work obsolete. Only five 5100 machines were sold, but one of the five is still in use producing, not labels, but boxes.[39]

INK SPOTS

"Yes, the inks. The inks came a long way, too. We had inks that we printed on all different materials and you'd have to have a different ink for almost every material." [40] Tom Wilson, of Salem Label, later with MPI, talked a lot about the problems the converters had with the inks available before the 1970s.

Some of those old inks would...dry so fast that you would hardly get any kind of a quality print....Back then earlier materials weren't treated so well for inks as they are now...the inks adhere to it [the material] better....If you were printing acetate, you had to have an acetate ink. If you printed foil you had to have a foil ink, or vinyl, you had to have a vinyl ink We used to print a lot of sticky tire tape, you had to have special inks.[41]

Lou Werneke knows about inks. "Printer's ink is the catalyst that is bound together by generations of Werneke's. My father was in the business, his father was in the business...and I have a son Matthew who started with me when he was fourteen years old."[42]

Werneke points out that what really got him in the water based and flexographic ink business was an oil ink project at Minnesota Mining:

> *One of the people I really liked over there was a...physicist by the name of George Ladda. I said [to him], "According to what you're telling me here—this was in '73, it was the first energy crunch—is that litho has reached its peak....web offset will reach its peak in a couple of years....and the two fastest growing printing processes are gravure and flexo." He says, "There have to be some changes made for the environment and for the energy crunches we have" ....and I decided I was going to dedicate some time and energy to develop a water based flexographic ink.*[43]

Werneke worked with a chemist at 3M.

> *We used to work every Saturday on water based inks. When we got to the point where it looked pretty good coming off of our press, we would leave on Friday night and would drive all the way to Winnipeg to Gibson Label. He had a couple of Webtrons...in his basement...and we would run all day Saturday, make some adjustments and see where we were and where we were going....When we decided we had a product that we could bring out....It was like a catastrophe.... You would call on fifteen people and they would [say], "You have a what? You have a water-based flexographic [ink]?" And what is so strange about the whole thing is that I did something about that, but 3M didn't.*[44]

Then Werneke began to look for better water soluble resins. A competitor was his biggest helper:

> *I had a friend in Chicago...Carl Billings....[His company is] one of our biggest competitors....He called me [about] a resin from Johnson Wax Company and it's soluble in alcohol, and under certain conditions it's soluble in water. It's the best thing I've ever seen."*[45]

So Werneke took fifty pounds of the experimental resin to a company called Pen Color Company. He had the resin made into chips and, "When I got those chips back to Minneapolis and we made an ink from it, it was the glossiest, most transparent thing you had ever seen."[46]

Now Werneke had the inks, but they still had problems with running the presses. He finally discovered that the residue of the solvent inks was contaminating the rollers for the water-based inks. They cleaned the presses and the inks worked.

But early water based inks were difficult to dry. So Werneke turned to the paint industry. "They have things...in all water-based paints....What you learned to do is crosslink the resins." [47] This allowed the paint to dry and to remain permanent.

Werneke believes that currently ninety percent of the TLMI narrow web members are into water based inks. The Louis O. Werneke Company, like MPI, Avery and others, has many divisions nationwide. Each is autonomous. Werneke's new company in the United Kingdom will be treated independently, as well. It makes it easier, says Werneke, "Just like the *Music Man*, you got to know the territory." [48]

LEARNING CURVE

FTA held its first printing seminar in New Orleans in 1972. TLMI had begun a Pressman Training Program earlier (in 1969), but everyone interviewed seems to agree that the seminar pioneer was Ed Vandenberg in Cincinnati, probably as early as 1963.

Ed Vandenberg and his salesman, John P. Jones...

> *promoted the very first seminars in the industry. The first was held at his plant...the second at the Elks Club...then they graduated to the Gibson Hotel. Of course they were promoting their equipment, but then they were also helping the industry. I [Jim Ferris] spoke as a rotary die maker....It was a good start [to what] now they call networking.*[49]

Marty Kesten remembered them as:

> *the only show in town. People used to come from all over hell and back to Cincinnati, which was not my idea of going to the nicest place....Took them into his manufacturing plant and showed off his press....They had an ink guy there and Bob Kennedy, who worked for Dennison...spoke on raw materials and adhesives....I remember going to that first one and finding it very interesting in two ways. One, that here was information being given out pretty freely...and two, that there were all these guys around that knew this stuff.*[50]

Today seminars are commonplace, even mentioned in advertisements for presses (Gallus is one).[51] DuPont and other chemical companies offer seminars, which are often sold out. Seminars have become a way to learn about the basics of converting and new techniques. Information is shared freely.

AND THEN, BAR CODES

Alan Campbell joined Avery/Fasson in Painesville, Ohio, in 1964, and one of his earliest assignments was

working with business forms converters in developing our earliest EDP products. That included working with the people that just produced blank pin-fed labels and also companies like Moore Business Forms....An early assignment was in developing a high speed splice....the adhesive would ooze through from the top and bottom of the splice and stick together and cause web breaks... A guy named Pete Hollish and I went and saw a company on the east coast called Label Art and they showed the kind of problems they had...and Sam Jackson developed the three layer splice. We called [it] high speed splice....[Today] it's a standard splice in the industry, I think.[52]

Variable information printing now represent fifty percent of the total market, according to Campbell. "That would be price marking, computer labels, thermal transfer, anything where you're putting variable information [from a computer] onto a label that may or may not be preprinted...It's the printing of variable information." [53]

Campbell remembered 1965 "when the EDP market was going to die...and then [again] it was going to die in 1973...then it took off." [54] Then the Universal Product Code (UPC) had a big impact, with literally thousands of UPC bar codes established.

Bar coding has gone well beyond [UPC codes for supermarkets]....[It's in] inventory systems, factories, routing systems where the bar code actually controls the logistics of the product as it goes down an assembly line. It's used to route baggage in an airport....We as a company control our entire inventory using bar codes with an integrated bar code system....We use bar codes to store quality information on products...identify when the product was made, what raw materials were made from what batch and from what supplier.[55]

Campbell pointed out that release capabilites have improved dramatically over the years, improvements in the base paper, improvements in consistency, improvements in moisturizing techniques.

> *Moisture has always played a critical roll that was not well understood...in terms of not only dimensional stability, but the strength of the paper. If the paper was very wet it would tear easily, and if it was very dry it would tear easily....But the release liner industry, as the volume increased, then [it] had to start responding....The biggest changes in advances of release have come from, one: improvements in the base papers, two: the use of solventless silicone...a friendlier system for label converting....We were the first, the United States, to get into the emulsion coating of silicones. [Pressure sensitive] finally made it in the middle seventies.*[56]

THE GREAT IMPETUS

In 1970 Bud Gray moved from Uarco to Fasson and in 1973 left Fasson to work for one of his former customers, York Tape and Label in York, Pennsylvania. (He later worked for Apollo Label and is currently at MPI.) Gray says that, back in the 1970s, only Avery and York were working on rotary screen printing as well as rotary hot stamping. However, Gray recalled:

> *One of the biggest upheavals in that industry at that time [about] 1975 was the gradual shift from rubber flexographic printing plates to photopolymer plates. When you printed on rubber you went from artwork to a magnesium engraving, to a Bakelite mold to a rubber printing plate. And with photopolymer plates you went from...artwork directly to a printing plate. We thought we'd be able to pass on significant savings to end users....The savings weren't there in dollars and cents, but it made for much faster makeready....The quality wasn't there at first. As in all innovations...the quality wasn't there. And you still had pressmen who absolutely refused to use photopolymer plates.*[57]

Gray stated that because of the market demand for higher quality labels, the industry was forced to make greater use of the improving polymer plates.

> *"We began to produce four color process work...but keep in mind that we all wanted to stay away from fleshtones."*[58]

Jim Hattemer of Graphic Resources agreed with Gray.

> *The photopolymer plate provided a consistent point of reference for the ink, press, and material manufacturers, against which they could work. [It] encouraged everyone to make their products and materials more consistently. It wasn't until the photopolymer plates' quality that you could go to primary labeling/decorating function. Along with that, the rest of the market could [also] grow.*[59]

INSTANT MARKET EXPANSION

One of the ways market growth continued was through the development of instant coupons. Hattemer said Graphic Resources patented some of the innovations and continued to develop the coupon business. In 1980 folded coupons began and enlarged the base. Now called *on product* promotional labels, these labels are an integral part of the business Hattemer developed.

Graphic Resources was sold to Engraph in 1991, and Hattemer pointed out that Engraph is

> *the largest player in the PSA Roll Label industry, over $100 million per year. Avery is larger, but that would include their base material group....Engraph, as a total corporation, is in excess of $200 million. [Engraph] also purchased Patton, Screen Graphics, Package Products and Blair.*[60]

"BEST COMPANY I EVER WORKED FOR"

Carl Williams had good thoughts about York Tape and Label, too:

> *To this day, [I believe] York Tape and Label was the best company I ever worked for....York at that time was an excellent converter. There was top quality, you could go back in the press room, go back in the art department, and it was as clean as your house, if not cleaner....York just had a culture within themselves....Everybody worked together. Everybody took a lot of pride in seeing the label come off the press and get die cut, you know, just like it should be.*[61]

Williams was sent to Cincinnati by York to maintain their business after the local manufacturers' representative, Andy Anderson, died ("right in his office"[62]).

We had Proctor and Gamble going, we had Cincinnati Microwave, we had Drackett, we had a company that makes Nutone....So I went out and started selling and really started to like it.[63]

From York, Williams moved to Dennison, where he remained. The Dennison story is in the next section.

IT ALL BEGAN IN A LOG CABIN

Stadia Colorado Corporation opened for business on April Fool's Day in 1970 in a log cabin on the outskirts of Denver, Colorado. Dan Mauritz and his wife and two children lived in the log cabin with a log warehouse behind it.

When Dan joined Stadia in 1972, it was a distributor of packaging equipment and did not print labels. Then Stadia sold a one color flexo press and supplied the customer with roll stock, printing plates and ink. "We edged into it," [64] said Mauritz:

Sometimes we would be talking to customers about printing their own labels. [Mauritz's previous experience with Bell and Howell had convinced him this was a wise choice.] They would say, "Well, I just don't want to get any ink on my hands. Why don't you print them for me?"[65]

The "print them yourself in-house" concept meant that the customers could get labels anytime they needed them. So, when Stadia started printing labels in response to customer demand they maintained the quick turnaround aspect of in-house production.

And we've stayed with that principle in our printing department. They are kind of proud of our fast turnaround....We measured business by dollars [invoiced]....Sometimes our competition measures how good business is by their backlog....We never quite understood that. Well, we understood it, but we never agreed with it.[66]

Their first labels were simple one color butt cut labels on pressure sensitive paper. In the first years Mauritz and his three or four salespeople would sell all day and print the labels at night. After a year of living in the log cabin, it was converted to an office and in 1979 they actually had a million dollar business in a log cabin!

300 MILES MORE OR LESS

Maynard Louis and Les Ordman claim that they started the concept of regional plants in the early 1970s.[67] Lord Label believed no plant could effectively serve a market more than 500 miles from the production center. As quoted earlier:

> *When you get beyond that there are too many good converters out there that just make you ineffective. So that's when we started the multiple plant deal. Now the sincerest form of flattery is to be copied and since we've done it MPI has done it and so has CCL.*[68]

MPI's Don McDaniel says he believes 300 miles to be the limit. His second plant was acquired in 1975.

> *A small label company called Lawson Label in Peoria, Illinois, came up for sale. We bought Lawson Label and the guy who was our plant manager moved to Peoria to run it.*[69]

In 1978 the Diagraph Bradley Company of Nashville, Tennessee, had a press for sale and MPI was interested. McDaniel and the Diagraph owner were discussing the price of the press when the owner said, "You know, the mood I'm in today, instead of selling the press, I feel like selling the whole damn company." [70] McDaniel bought the label division and sent one of his employees, "a single guy who didn't mind living in Tennessee,"[71] to run the two press operation.

> *[It] was in the basement of a drug store in downtown Nashville. When it would rain, the water would run in under the desk...so we got out of there pretty quickly, rented a building.*[72]

The *single guy* met a girl in Nashville, married her and decided to move back to Ohio. Bruce Jackson, an ex-Fasson salesman, was hired to run the operation, which, according to McDaniel is becoming one of MPI's better operations.

> *Then, in 1979, another Fasson, guy, Bill Estridge, who was regional sales manager...called me from Charlotte, [North Carolina], and said, "Don, have you ever thought of having a plant in Charlotte?"*[73]

Fasson was planning to move its sales office to Atlanta and Estridge didn't want to move. McDaniel and Estridge met halfway, in Charleston, West Virginia. The deal struck was that when Estridge had enough orders to justify a press, they would install one. A year or so later, a seven inch, six color Mark Andy was installed, a pressman hired, and the business firmly established. Unfortunately, the business didn't grow as rapidly as Estridge had hoped for and he soon left the company.

Who would run the operation? McDaniel remembered a small label shop in Akron called Ferrell Label. Owner Al Ferrell, who had worked for both MACtac and Kleen-Stik before starting his own business, had once done MPI a favor. MPI had received a large order for blank labels from the State of Ohio just when they were running full out and Ferrell had the available capacity on his one Webtron to subcontract the blank labels.

After Estridge's departure, McDaniel called Ferrell to say, "Al, why don't you take your Webtron, put it in your pickup truck, haul that thing down to Charlotte, and you run our plant for me?"[74]

So he did.

Between Estridge's departure and Ferrell's arrival, Fasson's Ed Mahnic, who sold MPI its first roll of PSA stock, ran the operation for six months, on his way to Florida.[75]

PHILATELY TO PRESSURE SENSITIVE

Duane Hillmer is another one of the recognized industry pioneers, starting out in the grease proof insert business in the late 1930s. It was in 1950 that his firm, Paramount Paper Products, started printing on 3M self wound tapes and gummed label paper. He did not go to pressure sensitive labels until the middle 1950s when he acquired a Chicago business called Metal Specialties Manufacturing Company. In the deal, Hillmer received three New Era presses and started printing pressure sensitive labels. In 1961 Paramount started coating its own self-adhesive materials.

Hillmer sold the business in 1965 to Nashua Corporation and worked for Nashua for a year and a half before retiring in 1967.

> *And I was very bored and I was only fifty years old when I retired. So I bought a stamp supply publishing business called the Scott Publishing Company. [While there] I had the opportunity to buy Epson Lithographic. That put me back into the label business...lithographic can and bottle labels. Then in 1976 I went out of the stamp world and bought my first Gallus press. I guess my claim to fame...on the current scene is I put the first Gallus R2500 rotary letterpress into the U.S.A.*[76]

ALMOST A DISASTER

He also installed the first electronic monitor Gallus sold in the United States. Today his firm, Epson Hillmer Graphics, is running six Gallus Presses. But it almost proved disastrous. After his first Gallus was up and running, Hillmer's biggest client, Safeway Stores, decided to produce its own labels!

> *And we had to do something and we started making what we call the curlicue label... which does away with the waste around the label. In other words it cuts through the backing paper linearly and the labels are separated with a cut cross-ways that does not go through the backing paper. So you peel them off and there's no waste. Curlicue got approved [to bid on the production of those tiny Day Glow postal stickers] and on the first bid there was one order for 715 million labels.*[77]

Hillmer is accustomed to new labels but is quite excited about a new label for a clear Lavoris product:

> *It's new to me....I never saw it before, decorating the backside of an informational label....you look through the front label, which is on film and carries not a hell of a lot more than a logo and some things about "best ever"...but anyway, you look through the liquid, you see this beautiful waterfall and if you happen to wiggle the bottle it looks like the water's really going. It's an exciting concept because...it opens the door to a much more decorative package.*[78]

He says one of the "highs" of the business "is the guys that work for me that are millionaires today."[79]

Some of those who worked at Paramount Paper and have remained in the business are Bob Smith of Avery, Joe Hyde who went on to form and sell three other businesses, and Sandy Burnstein of Nova Label in Washington.

A FAMILY AFFAIR

Another 3M tape salesman who quit to form his own printing on tape business in 1962 was Paul Myers, Sr. He and Lee Paul of Tape and Label Engineering were good friends. Dennis Hulton, Paul Myers' stepson, was working after school setting type and running a flat bed Markum letterpress. Hulton graduated from Millersville State College in Pennsylvania with an education degree, but always planned to work in the family business. He started full time in 1970.

The company, now Valley Forge Tape & Label, acquired a four inch, two color Mark Andy in 1963.

Even at that time they were doing some basic primary labels:

We were able to do some pretty nice work with the advent of the glass master plate from Graphanetics, were able to improve our print quality. We got into...solvent dye type inks...[which] were able to generate some nice strong, clear colors....In fact our first FTA...competition award was printed with glass master plates and dye type inks.[80]

In 1976 Paul Myers, Sr. died. Son Paul Myers, Jr. and Dennis Hulton became the owners.

It was intimidating, it was a challenge...for us to carry on with it. We had the advantage of basically growing up with the business so we knew the production end of it pretty much inside and out.[81]

Hulton credits Walt Webster of Flexcon with encouraging Valley Forge to go into the printing of films. Other suppliers have helped as well. "As far as rotary die cutting, I think Preston Engraving has always been a pioneer."[82]

Valley Forge Tape & Label considers itself a general roll label converter:

We've tried to align ourselves with more higher quality prime label...the four color process....We are a general roll label converter. We do a fair amount of industrial type labeling, name plate computer printable Mylar. We do a wide variety of...constructions [and] also serve a wide variety of markets.[83]

Valley Forge has experienced double digit growth for most of its years in business, as high as 25 to 30 percent, although it currently is running at what Hulton calls a more manageable ten to twelve percent. But he believes such growth will continue.

I would think it's going to continue, in part...[because] the printing process has improved the quality capabilities, the...quality has improved and continues to improve and is becoming [more and more] acceptable to more and more end users.[84]

CREDIT TO THE PEOPLE

It would be difficult to find many businesses that have been around since 1970 with the same four principals. Yet that's the story of Superior Label Systems. Three of the four partners were working at National Label, the fourth partner was a brother of one of the three. The brothers were Jim and Earl Gettelfinger; Tom Braig and Ken Kidd made up the foursome.

They sold bonds and bought a three-color Vandenberg press. Ken Kidd recalled:

> *We started out with a drum press. We moved into Mark Andy and Webtron a little later on down the road. The first label we ever ran was a little address label for my dentist. Probably the way I paid my bill then....We are to this day a major player in the dairy label industry...a lot of our business is in food labels.*[85]

But the real break came in the mid-1970s:

> *when plastics began to come of age. With the plastic and blow mold bottles of the dairy industry, and we got into...an effective...economical way of laminating [polypropylene] overlay for the dairy industry. Pressure sensitive labeling...with applicators [were] easy to put on and you didn't have to get into screening and glue labels and stuff like that. It began to compete with glue.*[86]

Superior has three divisions, a label printing operation, a label applicating division and a machine systems group that manufactures labelers. They offer a thermal transfer printer and see nothing but growth in thermal transfer.

> *The direct thermal seems to be used more in the short, fast applications such as the supermarket side. But thermal transfer seems to be the big hitter in the industrial market where variable information is printed and...the code must stay on the box, be grease resistant, be rub resistant...basically all that is thermal transfer.*[87]

Kidd said Superior's success is due to his

> *partners, our employees and our customers. That is our company. We've been blessed with good people and good customers and that's the name of the game....Of course our business went beyond our dreams. And that's a credit to our people—our employees and our customers. Without them you don't have anything.*[88]

INFINITELY MORE THAN INK ON PAPER

At Northeastern University in Boston in 1961 Paul Clifford was going to school fifteen weeks, then working fifteen weeks. Royal Label was where he worked, first in co-op, then part time, then full time. He ended up in sales service. He got the information he needed from the plant manager, the two foremen, and suppliers. The paper salesmen were helpful:

They were more crusader-type salesmen back in those days [early 1960s]. Today you don't see the mill guys making the calls like you used to. They were always there to ask questions. I'm talking like every week they'd be in, twice a week. Fasson guys. Morgan guys. 3M guys.[89]

Early on, most of Royal's PSARL work was in the food business. One or two colors and a "lot of Day Glow."[90] When Mrs. Moriarity, the wife of the owner, died in 1969, Mr. Moriarity wanted to sell the business to Clifford and suggested the young man find a partner. The partner was Bill Donovan, then Fasson regional sales manager in New England.

He jumped for the brass ring but good, so we sat down with Moriarity, the accountant, attorneys and we cut a deal. It was a very fair deal and we even paid him off before it was due.[91]

Bill Donovan remembered when he and Paul Clifford purchased Royal Label, then a half-million dollar company in New England. They started planning the day they took over the business, but "you never really know what the problems are until you own it."[92]

When I took it over they were deep in garden produce and supermarket labels....Saw very quickly the fallacy of that type of business....We can't make any money pushing garden fresh labels at 25 to 30 cents a thousand. I can remember making the decision one night to completely drop our biggest piece of business...$100,000 annually with a supermarket chain here in New England....It represented probably twenty percent of our business. Now I think quite honestly we didn't really have a choice. We weren't making any money. Machines were running like crazy and the customers were demanding as hell. They [would] call you at 11:00 and say, "By 3:00 I need labels." On that basis we really couldn't get into any other type of business and make any money.[93]

Six months later Royal Label knew it had made the correct decision. After a few so-so years, Clifford said it began to take off in 1973, with more industrial applications such as foams and tape.

We did a lot of medical work. We were making pressure sensitive applications [such as] EKG pads....We hired [Paul Ryan] a Fasson industrial man. He opened up some good markets for us. It's even more sophisticated now, we're into pharmaceutical work....We call a lot of people in the direct mail business, direct marketing, if you will. I think...five percent of our sales might be food related. [Now] we're calling on people like Gillette, Polaroid, Kodak, Hasbro, Reader's Digest... plus all the computer people located in New England.[94]

Royal Label became one of the pioneers in the use of the glass plate which Corning Glass Works had introduced in the middle 1960s. Although DuPont's Cyrel™ photopolymer plates soon made the glass plates obsolete, Donovan credited them as being a major contributor to improved quality.

Suppliers have always been excellent in the flexographic industry. Very willing to work with you to develop new inks, new plate materials, new tooling.[95]

When asked about the major developments in the industry, Clifford mentioned the DuPont Cyrel™ system and water based inks. He suggested the next breakthrough will be in post press development.

Bill Donovan recently sold his share of the business to Paul Clifford and retired at 62.

I'd run the course....I got a little tired of that 22 mile commute each way. Let me enjoy life. [But he'd do it all over again.] I thoroughly enjoyed the Fasson experience...a very important part of my education. I learned a hell of a lot, there's nothing like working at the home office.[96]

Donovan summed up his experience at Royal Label by saying, "It's a very vibrant, alive industry for the simple reason you're looking at something new and different every day of the week."[97]

Clifford, too, remains enthusiastic:

This is a very interesting business. It's a remarkably complicated business...[what's] the stock we're going to use, the adhesive you're going to use, [how will] the label be used, and printing it, die cutting it, converting it and the quality of it. It's a very creative business....Commercial printing, to over-simplify it, is purely ink on paper. But the pressure sensitive industry is infinitely more than ink on paper.[98]

ABE REICHMAN'S VICE

Al Spinello started talking to Abe Reichman about joining him at Blue Ribbon Label in August of 1973. It wasn't until February of 1974 that he finally decided, "to come over. I don't like to make rash decisions."[99]

Spinello had occasionally moonlighted for Reichman when Blue Ribbon was overloaded with work. Both men had experience with Ever Ready and at Alan Hollaender, but at different times. But Spinello said the move to Blue Ribbon was "the greatest thing that ever happened to me."[100]

> *[Abe] knew anything and everything about the New Eras. You could blindfold him and he could put that thing together, and just sitting here in the office and discussing various problems, he would hear something that wasn't correct outside and went right out...put an apron on and run the press. He was not one you could snow.*[101]

Abe Reichman died in December of 1990 and Spinello took over the leadership of Blue Ribbon.

> *He was 74. I tell you the honest truth, if it hadn't been that he passed away, he'd still be in it [the business] until they carried him out. This was his love....This was his vice. He loved it.*[102]

SCARED TO DEATH

Jim Smith apparently felt carrier landings were too tame, because he switched to submarine duty. When he left the service, the Naval Academy graduate moved to Anderson, South Carolina, and worked for Owens Corning Fiberglass for a number of years.

Then a friend whose father was in the printing business in Anderson recognized the need for PSA roll labels, and Smith and his friend moon lighted on the father's equipment. The two started Tag & Label Corporation in January of 1968, but the friend's father grew ill and the son had to take over the father's business...

> *and left me holding the bag...literally. I thought it was a good opportunity, but I was scared to death once he dropped out because I had no background in business or selling.*[103]

From the beginning Smith moved into important market niche applications, becoming an approved UL converter to service a number of area electrical component manufacturers. He also worked in the man-made fibers industry.

> *We produced labels and some other products of a tag nature. This is still our largest area of endeavor here. They have what is called "tracing." Tracing is a method [wherein the manufacturer] can determine the history of the product. We do a lot of bar codes... really what they are, are sequence numbers or sequenced bar codes.*[104]

"I WANTED TO KNOW MORE AND MORE"

Smith believed he was among the first label printers to become heavily involved with computers.

> *Part of this was due to my interest in the area and the fact that once I got involved in the business side of things, I began to take on more of an accounting... or financial posture than most people. I closed my own books, did my own taxes....I decided I wanted to know more and more. So that's the reason I got into computers early.*[105]

Tag & Label Corporation has just under one hundred employees, but 36 computer terminals.

> *We do everything on our system from accounting, automatic invoices, packing list generation, all those things. We have all the job specifications on computer. If we get a re-run we pull the job card out of the computer....I have six full time people here who are programmers and I'm one myself.*[106]

It would be a big surprise if Tag & Label Corporation were to win a TLMI label competition. They have never entered.

> *We don't do process color work. We've done a few process labels but I think that's something [you have] to do almost daily in order to stay up with it. We're more in the product identification field, our products are not very pretty, we try to do quality on them, but they aren't the kind of products [whose labels win awards].*[107]

It's another successful "find the niche" story in the PSARL industry.

EPA AND OSHA

Beginning with the Clean Air Act of 1970 which established the Environmental Protection Agency (EPA) and the establishment of OSHA (the Occupational Safety and Health Administration) in 1971, industry in the United States was faced with even more challenges. Firms using processes, products, or materials that harmed or had the potential to harm air quality, or materials or processes which adversely affected the health or safety of employees came under direct state and federal governmental scrutiny and regulation.

EPA and OSHA had a significant impact on pressure sensitive adhesive formulators and coaters, and to some extent, on the user of these materials, the PSA label printer. Until the 1970s solvents were the medium used for almost all adhesives, inks and coatings, and most effective solvents were quickly placed on the EPA and OSHA hit list because of adverse environmental effects and potential cancer-causing effects for humans.

Printers, coating machine operators, adhesive compounders—all began a mad scramble to comply with a new set of regulations forced upon the highly competitive business environment. Compliance with governmental regulations equates to higher cost of doing business. Non-compliance ultimately brings business failure.

Compliance took three basic forms:

1) *finding solvents which were not considered harmful by the EPA. Natural rubber formulations had already been phased out in favor of SBR, a synthetic rubber developed in World War II. Now the formula tions were adjusted for EPA requirements, keeping toluene level within the prescribed twenty percent limit.*[108]
2) *installation of equipment that would capture the solvent released and either recycle or incinerate it. In 1978, with EPA pressures on reducing solvent effluent, the silicone formulas were revised by many coaters to recover solvent.*
3) *development of new adhesives which needed no solvents in formulation or coating. Emulsion (water based) and hot melt adhesives were the result.*

Both water-based and hot melt adhesives are often more costly, but the savings on the use of solvents often offset those costs. Major increases in running speeds and improved safety (fire/explosion danger is minimized) are added benefits. Hot melt adhesives add a third advantage. Without solvent or water to drive off while curing the adhesive, there is no need for drying ovens. Not only does this lower equipment cost, but reduces the amount of plant space required to house equipment.

Manufacturers of the inks and coatings used in all forms of printing were impacted similarly to adhesive manufacturers. Water-based technology was the primary alternative to solvent-based systems. Out of this research evolved the ultraviolet (UV) families of coatings and inks which have proven to be a boon to the flexographic roll label printer. In addition to meeting EPA requirements, they are of superior quality than earlier products used by flexographic printers.

FROM ZERO TO $22,000,000 IN SEVEN YEARS

Bruce Motter, who had worked at Fasson and MACtac earlier, was one of the three founders of Chemtrol Adhesives, a company formed to take advantage of a new adhesive technology. The technology was acrylic emulsion. Production started in 1973 and when Motter sold his interest in Chemtrol in 1980 the company had sales of 22 million dollars!

Compac's Kleen-Stik was running emulsion pressure sensitive adhesives, having purchased a company in New York state which had developed emulsion-type PSA adhesives. Motter credits Compac as a true emulsion pioneer, but by this time they were no longer a key factor with the printing industry. So, when Chemtrol came "into the market it had very favorable economics on its side. MACtac was 100 percent solvents and Fasson was 98 percent solvents, and it allowed Chemtrol to very quickly gain a nice foothold in the roll label industry."[109]

A NEW CHALLENGE EVERY SINGLE DAY

In November of 1976 Jack Swartz joined Bertek (it was called Jonergin at the time). He had been with Kleen-Stik and lasted about a year with Compac before installing a coating operation in Mexico. He later worked on double coated tape for Bertek in New Jersey. He stayed with Bertek, running a coating plant, until he retired in September of 1991.

Bertek soon found a niche in pharmaceuticals, providing the transdermal nitroglycerine patch. Swartz said:

Those past fifteen years were very exciting because there was a new product every single day and a new challenge every single day. There was just a great deal of fun.[110]

Starting with one coater in a clean room, there are six currently. Water based adhesives account for about 85 percent of Bertek's operation.

We did that and went to water base and 100 percent solid silicones at the same time. The EPA came in and said, "You've got two years to clean up your act." About six months down the line we made the major switch on our major products. [111]

IT WAS JUST LUCK

Switching the adhesives to water base was simple, as Bertek purchased their adhesives and could change the specifications almost at once. The silicone proved more difficult. Swartz's equipment supplier was very busy and quoted a very hefty price and a 46-week delivery.

So I turned around and spoke to Dow (the only 100 percent solid silicone supplier at the time) and they gave me the basic idea over the phone—the basic concept of what you have to do, what type of controls—they were very helpful. And all over the phone. So I went ahead and designed it and put one together and it worked first shot out of the bag. It was just luck. No one does that.[112]

GREEN MOUNTAIN BOYS

You want me to tell you the best thing about it [Bertek]? The major people that the company grew up with were all local Vermont people. I've put plants in...New York and have operated them in Chicago and Los Angeles. I've put plants up in Canada, in England, in Mexico. The finest employees that I've ever seen in my whole life are the Vermont farm boys. They have a work ethic that has been lost in this country They are interested in what they are doing, they are interested in what their company is doing and they just take pride in what they are doing.[113]

SUPER SHARP

Preston Engraving's Jim Ferris says he believes "the industry came out of its infancy about fifteen years ago [about 1977]...and has matured to the point now where a person is not a pressman or presslady [but] an operator who's very knowledgeable. [They are] creating a product that's comparable with letterpress and offset litho."[114]

Ferris recalled the difficulty of cutting Mylar™ in the 1970s:

> *The die had to be super sharp. And then we had to educate customers that they could not use a Mylar die for cutting paper which may be full of titanium dioxide and other abrasive materials that would dull the die. The paper actually burst as it was being cut. Mylar would not burst, Mylar actually had to be cut....Some customers in the old days would take a die...use it on paper and then go back and they couldn't cut the Mylar.* [115]

Then the customers would yell at the die maker, the pressmaker and the Mylar supplier. Ferris said the problem was solved only when all of the above, plus the customer, got together and "solved the problem instead of pointing the finger."[116]

Ferris retired about 1985, but got the urge again and started a new business producing a new rotary die called the Red Devil. Later he sold that product to Chicago's Guard All and "retired all the way."[117]

Well, he still consults a bit.

CONSISTENCY, CONSISTENCY, CONSISTENCY

That's what Marty Kesten of Preston Engraving says are the three key characteristics that you have to work with.[118]

Kesten left the ailing aerospace industry and joined Preston Engraving in 1966 to work on dies for converters supplying labels for Kimball, Soabar and Monarch price-marking machines.

He remembered one special application that proved extremely difficult, the die cutting of adhesive bandage tapes. The adhesive was very gummy and stretched a lot.

"The over-laminating of labels became the next issue to challenge the industry," Kesten said. [119] This was probably in the early 1970s.

> *When they first came out glassine was still one of the liners of choice. Brittle, fragile—the adhesive would release from it all right, but you had to punch it, die cut it or perforate it, almost everything. It was a bitch for automatic application which was the thing that really made the industry grow.*[120]

By getting together with the base materials people they agreed to certain standards, but early standards were quite loose, plus or minus ten percent.

> *And if you're dealing with forty pound kraft liner that's about two and a half thousandths thick, that is a world of difference if you're trying to die cut to the liner without damaging it.*[121]

Kesten credits Dave Mages with creating a press with a movable anvil which allowed the pressman to adjust the cutting depth of the dies without touching the die itself. Kesten credited Fasson's Chuck Reed with the development of a step anvil roll and mentions Preston's own hydraulically-adjustable anvil.

Preston introduced a newsletter, perhaps in the late 1960s, that dealt with various converting problems (another example of disseminating technological know-how).

> *We had a series of newsletters and then we dropped it, and then we picked it up again and then most of our competition copied us. Fasson copied us.*[122]

Kesten credits Mark Andrews, Jr. with much service to the industry:

He's taken an outfit that used to make fingerprint presses, no two alike, and introduced mass production and standardization...through his company, much to his credit.[123]

THE COMPANY BARTERING BUILT

Thirty was the magic age two Corning executives had set for going into business for themselves. They didn't rush into it—far from it.

Gerry Gartner and Darrell Dochstader first met socially in the mid-sixties while both were working for Corning, and had even worked on a few projects together. Gartner, with a BS in physics, a masters in metallurgy and an MBA from Harvard, was working in market research and sales. Dochstader, who had a BS in engineering and an MBA from Penn State in marketing and finance, was transferred from corporate marketing development to control of sales for the glass modified plates Corning was making in Newton, New Jersey. Since Dochstader's father was in the dry cleaning business and Gartner's father was a small job printer, both had a kind of "entrepreneurial upbringing," according to Gartner.[124]

One of Dochstader's projects for Corning was to do a major study for Corning and the FTA (Flexographic Technical Association) regarding color process printing. The FTA published it, making the knowledge available to all. There were many test runs, conducted at various printing operations around the country.

> *Doug Tuttle, Jim Ely, and I spent the better part of three months going over all the test results, consumed many bottles of bourbon...then we had the first major...print color process dot study ever done. It became the bench mark standard.*[125]

It was to prove one of the key reasons that Gartner and Dochstader would choose to enter the flexographic printing business.

But how? They weren't exactly rich. The two each had $5000 to invest and the equivalent of Dorothy Durfee's $100 wouldn't go far in 1971. They could have opted to go in cheap, and print supermarket labels as so many successful flexo entrepreneurs had done, but they didn't want any part of that. Their decision was to enter the process color flexographic primary label PSARL business. It would take a lot more than $10,000.

IN THE WEE SMALL HOURS OF THE MORNING

Then they studied their chosen field thoroughly. Dochstader says:

We took a look at this industry, went to about twenty narrow web printers, got information from them—run speeds, equipment costs, prices [converters] were selling their stuff for, and all that, and in the middle of the night on Corning's IBM 360 we got a guy to run a cost analysis for us. We did about a fifty page analysis of this business and decided [in 1969] it was at the most early stages of development and very profitable.[126]

DARK SECRETS AT PHL

After signing confidentiality agreements, Mark Andrews, Jr., met first with Gerry Gartner in the Philadelphia airport.

Yes, Mark came in and Gerry Gartner came walking out and Mark said, "Gee, you're Gartner Printing [his father's firm] who I've been talking to all this time." Then Harry Mosher came walking in [Mosher worked with GarDoc for a few years] and Mark said, "My God, you're working with him, Gerry?" I [Dochstader] came walking out and Mark said, "I don't know what you guys are going to do but I'll support you."[127]

The problem was, of course, money. According to Dochstader, Mark Andrews agreed to make a number of engineering changes on his new press and, "We'd really support him with those changes and help him sell equipment if it worked."[128]

Andrews bought the idea. He aided the soon-to-be company by helping GarDoc get a loan to buy a Mark Andy for $40,000.

After the first year (1971) of operation GarDoc helped Wendell Dubbs develop a Typeflex distortion machine and Dubbs gave them one in return. Dubbs then got to publish that GarDoc was "using the stuff, and Dubbs sold machines all over and ended up selling his company and retiring."[129]

We needed a good ink system, so I did the product study for Inmont and I came up with a ten million dollar market. They told me I was crazy [but] figured a one million dollar market was still worth doing, and we figured out the pigments so everybody could mix their own inks....I told them we would test everything and help them market the product. Then, when they came to us with prices, I [met] with them and told them to double their prices.[130]

Dochstader says Inmont thought they were nuts, but they did raise the price fifty percent. "From then on, whenever we wanted something from Inmont we just called them up, because they said, 'We owe you big time.'"[131]

"So," says Gartner, "we formed the company in 1971 and quit our jobs and bought a six color, ten-inch Mark Andy press and moved to Southern New Hampshire because it was the place we wanted to live."[132]

> *We pulled up stakes, moved here, got the Mark Andy press, put on blue jeans and sweat shirts and started running this press. The first job we ran...was a four color process series of apples, peas, carrots and cherries. They [Mark Andy] were literally blown away with the quality of the work.*[133]

FIFTEEN OF FIFTEEN

Dr. William McGraw was, in Dochstader's words, "the father of the flexo plate at DuPont."[134] DuPont wanted to work with GarDoc after visiting their operation in New Hampshire.

> *We said, "We will work with you as a consultant and run your plates and help you develop a system, if you'll give us a system."* [135]

DuPont declined that relationship but did agree to enter into a consulting agreement which meant, as Dochstader said, "We worked out a deal where we'd be consulting and then we'd pay them the difference after a couple of years after we had some money."[136]

In 1973 or 1974, utilizing the DuPont photopolymer plate system, GarDoc entered fifteen categories in the FTA awards show and won fifteen, plus best of show. Overnight GarDoc rose from being an obscure start-up firm to a major player. DuPont publicized GarDoc and its domination of the FTA awards.

> *Over the first two years [after the show] I think we helped them sell many systems. This is how we did development differently. So we had the plate, we had the ink, we began developing the doctor blade systems with the new anilox rollers, and we were off and running.* [137]

"What kept us alive in the early years," said Gartner, "was not labels. It was shrink film for Corning's Hallmark Christmas tree decorations."[138] But from the very start, everything GarDoc did was process color.

COTTAGE FOR SALE

Even in 1978 the flexo business was still "primarily 85 line screen and our process work...was not really recommended," said Dale Bunnell, who joined Mark Andy that year as sales manager.[139] Bunnell's background was offset, and offset was selling to the larger markets. Flexo, Bunnell said,

> *...was still very much of a cottage industry. [What] makes the industry exciting is the fact that...it is filled with entrepreneurs. Smart businessmen who are out there making their own efforts pay off....They would buy a small press....It would often be a man and his wife and he would sell during the daytime and she would keep the books and he would run the press at night, and within a year...he would be ready for another employee or two and...they would end up buying a press every 24 months. [That's how] many of these companies today that are in the ten to twenty million dollar range got their start.*[140]

FOUR HOUR PAY OFF

One type of application that no one could have imagined in the early days of pressure sensitive adhesive labels is the transdermal.

The best known transdermals are the seasickness patches. If you took a cruise anytime in the late 1980s or in the 1990s you would be convinced that the Earth had been taken over by Martians, each wearing a funny-looking patch behind the ear.

The original transdermals were pressure sensitive materials with a hypo-allergenic adhesive and a *well* in the center. The user would fill the well with medication, such as nitroglycerine. Applying the pressure sensitive material forced the medication against the skin, which absorbed the drug. Today's transdermal technology includes compounding the medicant directly into the adhesive.

Bunnell said the owner of the first Mark Andy made especially to produce transdermals had his investment returned in the first four hours of production![141]

THAT'S ANDY, ALL OVER

When Bunnell joined Mark Andy in 1978 the man he replaced was Ray Kuntz. Kuntz predicted a none-too-rosy future for his replacement. He said that:

> *I [Bunnell] was coming into an industry where basically it had peaked out, that he had sold flexographic presses to probably everybody out there who could use it. He... specifically said that the industry had saturated. I said, "Ray, I don't believe that's the way it is. It looks to me there's an awful lot of opportunities."* [142]

Bunnell proved to be correct. Companies were buying a press every twelve to 24 months, plus there was a constant stream of new companies into the business. Mark Andy continued to grow.

UNSUNG HEROINES

Probably the hardest workers and those with the least recognition are the wives and children of the entrepreneurs who started the PSARL businesses.

> *My wife has been involved in the company all that time. Actually 23 years....My oldest daughter is a kindergarten teacher, my next daughter is graduating from Colorado University this year. Then my son is sixteen, so I'm expecting one of these two, maybe both of them to join me. (Dan Mauritz, Stadia Colorado)*[143]

> *We both had small families at the time....We knew how to get by on a very little amount of money. (Gerry Gartner, GarDoc)*[144]

> *I was actually real fortunate. I started when I was a kid....in our garage...we used to do bumper stickers back in the early '50s....So I started silk screening when I was about eleven or twelve years old. (George Noah, Lewis Label Products)*[146]

> *He [my father] mailed out samples to a mailing list that he acquired....His family and children were putting samples in*

envelopes in the basement....I would work medical conventions in the summer time and I ran a press and those kinds of things. So I was basically weaned in the business....I probably would have started working for someone else...instead of being the boss's son...and the tough expectations of the father might have been mellowed out a little bit. (Jerry Nerad, Timemed)[147]

Sometimes the wife did even more than keep the books. Often she was the sole source of income until the company began to show a profit.

[My grandfather] started doing work for some of the local merchants at the time, including some of the bootleg liquor labels....My grandmother made the bottles and the stuff that went into the bottles. He just made the labels. (Dick Capuzzo, General Trademark)[148]

We lived on what my mom made as a secretary....We had some tough times the first couple of years....It's tough to grow in that environment....I have worked since early on, often for nothing—[no] child labor law, by the way....My mom, Rachel, is probably the only one who's attended all the boring meetings at TLMI. She's not just a spouse that comes along for the ride. She's his business partner. (Andy Beck, API Graphics)[149]

My wife at the time was a school teacher. Found a small machine shop near Toronto and we put, I think $5,000 down between us....My wife kept us alive for that period of time. (Allan Prittie, Arpeco)[150]

My wife and I went down there [North Carolina] and we both worked for ten weeks....[My son] Tom runs a press at MPI. [151] *(Jim Wilson, MPI)*

My wife and I sat down and I said, "What are we going to do?" [She said,] "Well, why don't we just go for it." We were living on her salary, [I was] driving her to work and taking her car, too, while I was on the road, and picking her up after work. We had no money. (Dan Tomlinson, Labelcraft Products)[152]

Not all the wives were happy:

My wife absolutely hates to shop with me....I get out into La-La Land. I'm out there...looking [at labels], gee, I never thought of that. I can't believe this is being done.....I start to think where can I go to sell this. [153] *(Mark Wert, MPI)*

After spending almost five years in Painesville, Ohio, my wife packing the bags every day and I was unpacking them every night...I said this can't go on forever. And I said I have no choice but to go back to New England where my family's happy. [154] *(Bill Donovan, Royal Label)*

[Klein: Do you have any regrets to coming into the pressure sensitive area?] Only my wife...Kitty. [but] she has been as active as I have been but she's been more on office work and stuff....[Klein: Is she still active?] She comes occasionally. She comes in to help out. [My son is running the business.] It's basically his business now. (Harold Sidrane, K. Sidrane, Inc.)[155]

[My wife] said, "This is our life, our income?" We had a couple of kids in college at the time; it was kind of a low point in her life....It took at least three years [until she became more comfortable]. (Don Buchta, Mid America)[156]

Children worked as well, after school, on Saturdays and during vacations. Sometimes it was the only way they could see their parents, since 100 hour work weeks appear to have been common.

I was born in 1939 so I had the advantage of growing up in that environment...hearing a steady diet of rotary printing, and printing on tape...it's been a lifelong thing, from the age of seven years old. (Mark Andrews, Jr., Mark Andy)[157]

I came down here with my father...I started work in the shop as the person rolling Christmas tape...I was seventeen years old. (John Garber, Label Technology)[158]

I worked in the family business summers, during the years I was in school. In fact, I dropped out of school for a year to work in the business full-time and then went back. (Dennis Hulton, Valley Forge Tape & Label)[159]

Father started a label business in 1957, [I] worked in the business after school and in the summer. (Jim Hattemer, Graphic Resources)[160]

I started working here part-time while in school as early as '52-'53. And I joined the company full-time in '58....I'm really the infamous third generation. (Bill Muir, Grand Rapids Label)[161]

My father was in the business, his father was in the business, and I've devoted my entire career to the same industry....And I have a son Matthew who started with me when he was fourteen years old and decided that this was the thing he wanted to do all of his life. He started in the plant we have in California; he's back here now. (Lou Werneke, Louis O. Werneke Co.)[162]

My son is president. My grandson just joined us. He was in the air force and he has finished his duties there and he's going to come here the first of January. (Henry Anderson, Salem Label)[163]

My family, of course, has been a large part of my driving force to continue to achieve success....I have three sons in the business with me and my wife and my brother-in-law. (Bert Kennedy, Kennedy Group)[164]

[My son] is now president of the company. He went all through the hard and dirty work. His mother thought his finger nails would never get clean. He had to sort of hide them at the dinner table, you know, turn your knuckles under. (Percival Wise, Wise Tag & Label)[165]

DOWNSIZE TO SURVIVE

Columbus, Georgia native Jim Hart had majored in accounting, but drifted into sales in a number of packaging firms before becoming product manager for Monsanto's clear plastic meat trays. In 1969 his boss and good friend, John Jacobsen, purchased the controlling interest in Allied Gear Machine Company.

At that time they were failing. We kind of got into the business because we saw it as a terrific growth opportunity. We'd been extremely successful in selling and marketing things for major companies....We saw labels as being an opportunity in that area.[166]

Allied Gear manufactured a series of presses. Hart said they were "pretty good presses"[167] but the company lacked the money and the background to market them successfully. So Jacobsen and Hart decided to cut the line to a single one color model.

[We] downsized in order to survive and we grew bottom up...funds generated out of profits. I really started by selling the one press. I was delivering it out of the back of my station wagon. The first one I sold was to David Higgins of American Design. It was $2,995.[168]

Gradually the cutting dies became more important than the presses themselves, but selling both created some good selling opportunities.

We had Ed Hero in Cincinnati in the 70's for a die customer and they bought some presses from us. We sold Monarch Marking and Lord Label was big even in those days....Because we did both the presses and the cutting dies, we got involved in many more converting projects.[169]

Hart remembered converters asking him how they could overprint an acceptable bar code. He believes the label industry is really a "series of specialized markets."[170]

[There] are maybe ten bar code label manufacturers Then you'll find another group of people strictly service the retail pharmacy market....Then you get another group [such as] Moore [who deal with] variable information. If a guy comes to me with a project and says, "Can you do this?" we say, "Yes, here's the machine that will do this" ....We don't take that project, put it in a magazine or take it to other customers and say, "Gee, look what we did for this guy." But if another guy comes to us with the same project, then we'll say, "Yes, we can do it." [171]

LEVELING THE PLAYING FIELD

Hart credited all suppliers, not just press manufacturers, with leveling the technology. Organizations such as FTA and TLMI are valuable sources of information. "Say a guy in New York is a friend of a guy in Florida and they're not competitive and they share information."[172]

Hart said when they entered the business about 1970 there were three die manufacturers. He estimates there are eighteen today. And he pointed out that die costs are actually slightly lower.

He also said that the PSARL industry is into product manufacturing! Johnson's Odor Eaters™ are being produced and silk-screened on a rotary press, producing excellent cost savings for the manufacturer. Hart is also bullish on America.

> *Certainly we've lost a lot of markets, but this whole thing [the PSARL industry] speaks to competitiveness and to ways [we can] do something better.*[173]

THE TEACHER WHO DIDN'T WANT TO RETIRE

Bert Kennedy of the Kennedy Group, Inc., went to Bowling Green State University in Ohio and became a teacher and a football, basketball and baseball coach. He quit to sell industrial tape and dispensers for 3M. Why?

> *All I heard [at school] was how long before we retire.... After six months I had upset stomachs every day. After I left I'd see guys [his former fellow teachers] and they'd say, "You did the right thing. I got eleven more years to retire." And I'd say, "Oh, my God, you're only forty and you're already counting yourself out."* [174]

At 3M Kennedy helped engineer dispensing equipment. After three years he joined a paper distributor for two years, then worked for a converter (Excelsior Tape) for two years, selling pressure sensitive tape and printed labels. He left under friendly terms in 1974 to set up the Kennedy Tape and Label Company. He started with 1500 square feet, one Mark Andy 810 and two employees.

Kennedy's firm almost failed on its first order. The material would not strip. The supplier's sales manager had quit, there was no salesman in the territory and the technical man was in the hospital.

So we [had] paid C.O.D. due to no credit history, which was my last $10,000, and, on top of it all, my new machine broke down with electrical problems. So that was my welcome to the business. I had mortgaged my house, I had mortgaged everything I had and I had a wife and three children. So there wasn't any more to do other than just keep going.[175]

The first six years were difficult. "If someone owed me a thousand dollars, it was like, 'Oh boy, what are we going to do now?'"[176] He had started his company during the 1974 recession and when the recession struck again:

In '79 to '81 I would say were the longest two years of my life, because that was the other recession and it was up in the 22 percent interest rates...and at that time I expanded my business....I've done things a little unorthodox.[177]

He expanded by buying three presses and two rewinders, and the timing was just right. Kennedy says he doubled his business in two years!

It was the CB (Citizens' Band) radio market that really got Kennedy going. An account he had served while working for 3M went from six to sixty million dollars a year.

We labeled all the CB antennas, all the packaging, all the cartons, die cut foam, the double coated tapes...probably made over one hundred items for that company alone. We became pretty much label experts with CB manufacturers which grew from a half a dozen to more than forty by 1977. The one I never got to was Radio Shack....We were just too small. I was in Texas then, so I was thinking big. I had a big company background, but I had this small company wallet.[178]

YOU CAN'T JUST GET BY ANYMORE

"If you talk to bankers who deal with small businesses," said Kennedy, "you will learn that 85 percent of [the small business owners] go to work with no plan everyday."[179]

I think in our industry it has gotten so technical that you have to know how to make a plate. You have to know what kind of sticky back to use. You have to know what kind of inks, what anilox roll to use, the metering system, what doctor blade, what you are going to do with your recycling system, you've got to know what your costs are, if you're buying right, if it's the right adhesive, the right liner, the right automatic labeler. You can't just get by any more.[180]

Kennedy must be a good boss. Four of his employees have been with him from almost the very beginning.

My family, of course, has been a large part of my driving force to continue to achieve success....I have three sons in the business with me and my wife. [My] brother-in-law has been with me eleven years.[181]

In spite of the fact that the Kennedy Group is successful, with fifty employees in a 35,000 square foot location, Kennedy says he is "not a manager-type person. I like creating things, and making things grow, and building things."[182]

RECESSION-PROOF?

The oil crisis of the mid-1970s and the accompanying world-wide recession altered the growth curve of many PSARL companies. All industries which relied on petrochemicals were affected by spiraling petrochemical prices and the six-month long oil boycott that began in October of 1973. "All of the petrochemicals that we used shot up in cost...the solvents that we had depended on tripled in cost over three years,"[183] said Chuck Miller of Avery. The cost of the base stock grew dramatically, as did the costs of release liners and inks. This meant that prices to converters needed to rise if the suppliers were to continue making the profits they had become used to.

But the 1960s and early 1970s had been a period of explosive growth, and many companies found themselves with increased plant capacity, improved equipment with greater volume, and new competitors in the pressure sensitive field. This created a squeeze on the suppliers. Their costs rose fifteen to twenty percent but they felt they could pass on only five percent or so. Profits plunged. Some companies, including Compac's Kleen-Stik, failed. The early 1970s were cruel times for PSARL suppliers.

During the recession, however, most converters (in effect, subsidized by their suppliers) continued to add to their numbers and profits. Double digit increases for converters were the rule, not the exception. It was the materials producers that were the hardest hit. Their units sales rocketed upward, but their profit margins fell sharply.

Avery felt the blow. 1975 was the first year since 1935 that the company had not grown. Sales fell to $282,608,000 from $296,348,000 in 1974. Net income fell even more dramatically, from $16,914,000 to $5,016,000.[184]

The recession hit Fasson Europe particularly hard. Prior to 1975, Fasson Europe had been producing sixty percent of the company's profits. Europe was even harder hit by the world-wide recession than was the United States, and Fasson Europe was facing new pressure sensitive competitors who were lean and hungry. It took almost ten years for Fasson Europe to recover from the recession.

AVERY INTERNATIONAL CORPORATION

Although Avery had established European franchises in the 1950s and an International Division in 1961, the company name was officially changed in 1976 to Avery International Corporation, reflecting the importance of the overseas operations on the company as a whole.

But before the name change there were problems to be solved. For example, a new Avery Label plant near Glasgow, Scotland was not producing well:

> *The methods that they used in Scotland were dated about 1890. They were running the presses at a quarter of the speed that we ran them in Monrovia. We lubricated the presses at the start of every shift, and then ran them like the dickens. In Scotland, they only lubricated the presses every two weeks. They were running the presses slowly to save on wear and to save on grease. It was very frustrating.*[185]

John Arozena finally taught the Scottish foremen that, in this case, at least, the American ways worked best, and the plant prospered.

Other problems included ethics, particularly the concept of bribes to get work completed. Although the bribe is commonplace in some countries, Avery refused to comply. "The money paid is not the issue," said Stan Avery. "It's just not good business. You can't have one part of company operating in a corrupt manner, and the rest not."[186] It took time, but the Avery emphasis on honesty, initiative and creativity became the emphasis of the overseas companies as well.

Another problem was provincialism. Avery thought of Europe as Europe, not as separate countries. The local companies' attitudes were parochial—Germans worked and sold to Germans; English, to English, and so forth. The Avery subsidiaries were forced to change their ways of working and selling. In some cases, such as the Swedish Fasson company, the sales manager refused to set the prices that Fasson required, so the director of Fasson Europe had him replaced. Other franchise owners were bought out. "By 1969 Fasson Europe accounted for two-thirds of the profits earned by the entire corporation."[187]

Fasson Europe then expanded into other parts of the world, Australia (1971), Brazil (1976), and South Africa. The U.S. Fasson company started plants in Mexico (in the late 1960s), Japan (a joint venture begun by Bill Klein in 1971) and Korea (the most recent plant).

THE SEVEN YEAR COMPANY

The soaring price of petrochemicals caused the breakup of Compac Corporation. Compac had been created in 1968 by Garrison Brinton and John Dickenson (of Ludlow Corporation) as a holding company to purchase companies in the coating business. They acquired Kleen-Stik from National Starch and kept the Industrial Tape Division of Ludlow.

Dick Riley joined Brinton in 1970, building base material plants in Mammoth Junction, New Jersey and Fort Lecia in Ireland. The base materials business was later sold to Essex Chemical of New York.

Seven years later, in 1975, Compac was no more, but Label-Aire, the automatic labeling division of the Compac group, was still a strong, viable company, so Riley and Brinton kept the label applicator business.

> *We had put a quasi-distributor program in place while it was [still] part of Compac. But we recognized that if we could develop a group of distributors...who manufacture labels, who had a high degree of professionalism and were willing to invest the capital that was necessary to make labels and select sales forces who were trying to distinguish themselves—not just label printers, but systems people.*[188]

They selected approximately twenty distributors and formed contractual relationships in 1976 with them for non-exclusive rights to sell the products. The program began in 1975 and is still Label-Aire's distribution system today, currently with fifteen such companies. Label-Aire works closely with its distributors on automatic labeling systems. The big selling point?

> *[It's] the ability to take that applicator and place it in the product line...the angle of the machine entry, nose down, nose up, wherever you can get it into your product line, we have the orientation already existing. It's just a matter of literally rolling the head up and put sensors on it and start labeling.*[189]

Label-Aire is a highly successful company, with compound double digit growth over the seventeen years Riley has been with the business. Riley believed, "We're the [largest] supplier of applicators in the United States for automatic applications."[190] He sees new opportunities for the PSARL industry with recycling and other environmental concerns.

They are also working for improved quality in both design and manufacturing. As 100 percent computer assisted design (CAD) company, Label-Aire is today:

> *embarking on ISO 9000 certification. We are hoping to acquire it within the next twelve to eighteen months. We have a joint venture partner in Denmark...who is already ISO 9000 approved, our joint venture operation will be from day one [running] under ISO 9000. We have to go to ISO 9000 across the board in the states....The companies that don't do that in the next few years, it's going to impact their [business] in the United States as well as Europe....We're going to have a global economy and these companies are interlinked worldwide.*[191]

BANKRUPTCY PAYS

The recession may have hurt some companies, but it certainly helped Joe Barnette of KAPCO (Kent Adhesive Products Company in Kent, Ohio). Barnette worked for MACtac for eight years before joining MPI. He remembered when Don McDaniel was going into business he stopped in on Barnette on his way back home after purchasing his first machine. It was in McDaniel's car, a three-inch Mark Andy.

At MACtac Barnette had worked with Tom Pace, Gordon McMann, John Brannigan, Charlie Brusso, Russ Paddock and Leon Kaston, all among the earliest salesmen of pressure sensitive materials. In about 1974, after three years or so with MPI, Barnette and a man named Bob Sibley started KAPCO. Sibley died only five years later.

Of all the ways to get started in a business, KAPCO's origin may be the most unique:

We ended up buying a company called Carr Adhesives [which produced pressure sensitive consumer products] that MACtac had taken over because they had gone bankrupt. MACtac had in turn attached it onto Elgin Label [another bankruptcy]. A silk screener had bought the whole shooting match but wanted nothing to do with the consumer product....One day [the man running Elgin Label and Carr Adhesive for MACtac] called me and said, "Do you want to buy this stuff?" [192]

The price was one thousand dollars and Barnette agreed.

Here we are, we had this little plant with nothing in it except the slitter and two presses. Here came all this material and we start going through it and found a stack of purchase orders that had not been filled. So, within the first three or four weeks we filled like four thousand dollars worth of purchase orders....Some of those purchase orders were six months, nine months old and we figured, what the heck we had nothing to lose. So we shipped them....It worked out great for us.[193]

FINDING INTELLIGENT LIFE

In 1979 Larry Kunkle decided he had had enough of corporate ways, so with a partner, he founded Heritage Graphics. The company, said Kunkle, "started selling not only pressure sensitive labels, but other packaging things such as tags, folded cartons, process printed corrugated boxes...a wide variety of things."[194]

To begin at the beginning, Kunkle left the U.S. Army in the mid

1960s and spent three years selling corrugated packaging for Container Corporation of America. Avery Label Systems offered him a position in Louisville which he accepted in 1969. He says his territory grew from zero to about a million in his ten years with the firm.

Much of his business was with the huge General Electric Appliance Park, where he helped install hot stamping equipment for serial number plates. Later he sold a lot of pin feed pressure sensitive labels.

Although most of his time was spent in Kentucky, Kunkle was headquartered in Avery's Cleveland plant, whose manager was Lynn Barden. Kunkle credited Avery for providing "excellent sales training, excellent product backup, product knowledge. You knew your product and I think people respected that."[195]

However, Kunkle also remembered tremendous turnover in the sales force. Why?

> *There were a lot of egos in Avery. The product managers and sales managers were paid according to a...scale established by upper management. And the salesmen could—if they had several good years--make more than managerial people....One particular year that I distinctly recall—'76—I made more money than Bob Fisher did.*[196]

Robert Fisher was Group Manager of Avery Label. So Kunkle, like Bill Eiseman at Mark Andy and George Collons at Kleen-Stik, decided to strike out on his own.

> *They had a lot of little tricks to lower your salary....And they expected you to sell more....The last quota adjustment really did me in. I had several friends who had left Avery and gone out on their own as manufacturer's agents and reps, so that's what I elected to do.*[197]

Today Larry Kunkle, who represents MPI in Kentucky and Indiana, says two-thirds of his business is pressure sensitive. He sees great opportunities in instant redemption couponing, in holography and thermal transfer:

> *Couponing is really big....The instant redeeming style coupon is a big growth item. Holograms are going to be a big growth item....MPI [has] developed a hologram that has a bar code you can engrave in the hologram and it's only scannable by certain kinds of scanners, so it's virtually tamperproof.*[198]

One area that is using the tamperproof hologram in a big way is the compact disc industry, which is so vulnerable to counterfeiting. Kunkle is really excited about thermal transfer because it allows him to achieve his most important job, "to help the customer save money by lowering his total applied cost." [199]

The cost per thermal transfer is coming down. Thermal transfer ribbon and machines...[are] state-of-the-art right now for in-house label imprinting. More people...can see it costs money to set up an item in inventory. You're going to have to order 200 different bar code labels....It takes somebody's time to keep track of [them]. If you can order a blank label, have a PC sitting there and print your own labels in-house, you've eliminated a lot of indirect costs. We push these things to intelligent people. If they can't see it, we go on and find an intelligent life somewhere [else].[200]

CHRONOLOGY OF APPLICATIONS, 1970-1980

1971	Decorative/Personal Care Labels (Ceramatic)
	Variable Information Weight Scale Labels (Plain Paper)
	Price Marking Gum Labels
1973	Industrial Bar Code Labels
	Automotive/Commercial Bar Code Labels
1974	Milk Jug Labels
	UPC Labels
	IRS Forms Labels (piggy-back)
	Tamper-Proof Labels
	Pressure Sensitive Postage Stamps (paper)
1975	Primary Flexographic Process Color Labels
	Magnetic Tape Labels

CHRONOLOGY OF INVENTIONS/ INNOVATIONS, 1970-1980

1970	Water Based Pressure-Sensitive Adhesives
1971	Water Based Flexographic Inks
	Ceramic Coated Anilox Rolls
1972	Electronic Feedback Tension Controls
	UV Rotary Letterpress
1974	Photopolymeric Plates
1975	Rotary Hot Stamping
1976	Electronic Controls for Adhesive Coating

THE MAN WHO WOULDN'T QUIT, PART FIVE

In 1975 Harold Scherer was on vacation and decided to drop in and see his friend Leo Middlebrook of Compac's Chicago office.

> *I walked in to see [the man] I had hired many years ago and he gave me a kind of funny look as if I should know something. So he said to me, "By the way, when did you leave for vacation?" I told him. "Maybe you don't know what's going on now." He said the company went broke and the bank closed us up. So in the meantime I continued my vacation and while I was away I made a couple of contacts and when I came back home I went to work for Westcote, Inc.*[201]

Westcote had been formed by Mort Allerdice and Angelo Liccardi who had originally worked for Avery. Westcote was not a manufacturer, but a distributor. When Scherer joined them total sales were $800,000. By 1987 Scherer's own sales alone were $2,500,000![202]

[1]Clark. The First Fifty Years, p. 132

[2]Andrews, Jr., interview p. 25, 26.

[3]*Ibid.*, p. 27.

[4]*Ibid.*, p. 35.

[5]Norman manuscript, p. 18.

[6]Ferdinand Rüesch interview with Bill Klein, p. 2.

[7]*Ibid.*, p. 4.

[8]*Ibid.*, p. 5.

[9]*Ibid.*, p. 9.

[10]Excerpt from Gallus History manuscript by Beda Künzle.

[11]Mark Herrman interview with Bill Klein, p. 9.

[12]*Ibid.*, p. 10, 11.

[13]*Ibid.*, p. 17, 18.

[14]*Ibid.*, p. 22.

[15]*Ibid.*, p. 33, 39.

[16]Webster interview, p. 14.

[17]*Ibid.*, p. 17.

[18]*Ibid.*, p. 19.

[19]George Noah interview with Bill Klein, p. 3.

[20]*Ibid.*, p. 3, 4.

[21]*Ibid.*, p. 4, 5.

[22]Bob Smith interview with Bill Klein, p. 1, 2.

[23]*Ibid.*, p. 5.

[24]*Ibid.*, p. 13.

[25]*Ibid.*, p. 23, 24.

[26]Norman manuscript, p. 13, 14, 15.

[27]Eagon interview, p. 15, 16.

[28]*Ibid.*, p. 17.

[29]*Ibid.*, p. 19.

[30]*Ibid.*

[31]*Ibid.*, p. 21.

[32]*Ibid.*, p. 29.

[33]*Ibid.*, p. 33, 34.

[34]Estridge interview, p. 12, 13.

[35]*Ibid.*, p. 14.

[36]Peter Menzian and Tricia Basto interview with Bill Klein, p. 4.

[37]*Ibid.*, p. 4.

[38]*Ibid.*, p. 10, 11.

[39]Garber interview, p. 4, 5.

[40]Tom Wilson interview, p. 10.

[41]*Ibid.*, p. 10, 11.

[42]Lou Werneke interview with Bill Klein, p. 1.

[43]*Ibid.*, p. 3, 4.

[44]*Ibid.*, p. 7, 17.

[45]*Ibid.*, p. 9.

[46]*Ibid.*

[47]*Ibid.*, p. 22.

[48]*Ibid.*, p. 14.

[49]Ferris interview, p. 11.

[50]Marty Kesten interview with Bill Klein, p. 17.

[51]TLMI, Gallus™ advertisment.

[52]Campbell interview, p. 2, 5.

[53]*Ibid.*, p. 3.

[54]*Ibid.*, p. 4.

[55]*Ibid.*, p. 5.

[56]*Ibid.*, p. 6, 7, 8, 18.

[57]Bud Gray interview, p. 10.

[58]*Ibid.*, p. 17.

[59]Jim Hattemer letter to Bill Klein, p. 2, 3.

[60]*Ibid.*, p. 3.

[61]Carl Williams interview with Bill Klein, p. 5.

[62]*Ibid.*, p. 6.

[63]*Ibid.*

[64]Dan Mauritz interview with Bill Klein, p. 7.

[65]*Ibid.*

[66]*Ibid.*

[67]Avery started regional offices earlier.

[68]Ordman interview, p. 19.

[69]McDaniel interview, p. 23.

[70]*Ibid.*, p. 24.

[71]*Ibid.*

[72]*Ibid.*

[73]*Ibid.*, p. 25.

[74]*Ibid.*, p. 26.

[75]Mahnic interview, p. 24.

[76]Hillmer interview, p. 6.

[77]*Ibid.*, p. 13.

[78]*Ibid.*, p. 15, 16.

[79]*Ibid.*, p. 10.

[80]Dennis Hulton interview with Bill Klein, p. 4.

[81]*Ibid.*, p. 12.

[82]*Ibid.*

[83]*Ibid.*, p. 14.

[84]*Ibid.*, p. 19.

[85]Ken Kidd interview with Bill Klein, p. 2.

[86]*Ibid.*, p. 4.

[87]*Ibid.*, p. 16.

[88]*Ibid.*, p. 11.

[89]Paul Clifford interview with Bill Klein, p. 2.

[90]*Ibid.*, p. 3.

[91]*Ibid.*, p. 4.

[92]Donovan interview, p. 13.

[93]*Ibid.*, p. 15.

[94]Clifford interview, p. 5.

[95]Donovan interview, p. 16.

[96]*Ibid.*, p. 22, 23.

[97]*Ibid.*, p. 19.

[98]Clifford interview, p. 9, 10.

[99]Al Spinello interview with Bill Klein, p. 4.

[100]*Ibid.*, p. 5.

[101]*Ibid.*

[102]*Ibid.*, p. 15.

[103]Jim Smith interview with Bill Klein, p. 1.

[104]*Ibid.*, p. 3.

[105]*Ibid.*, p. 4.

[106]*Ibid.*, p. 6.

[107]*Ibid.*, p. 7.

[108]Norman manuscript, p. 11.

[109]Motter interview, p. 9.

[110]Jack Swartz interview with Bill Klein, p. 7.

[111]*Ibid.*, p. 8.

[112]*Ibid.*, p. 9.

[113]*Ibid.*, p. 19, 20.

[114]Ferris interview, p. 19.

[115]*Ibid.*, p. 17.

[116]*Ibid.*

[117]*Ibid.*, p. 21.

[118]Kesten interview, p.7.

[119]*Ibid.*, p. 5.

[120]*Ibid.*, p. 5, 6.

[121]*Ibid.*, p. 8.

[122]*Ibid.*, p. 18. *Fasson Facts* eventually became an extremely useful publication, but the first ones put out by Fasson were essentially puffery. (Klein).

[123]*Ibid.*, p. 21.

[124]Gartner interview, p. 1.

[125]Darrell Dochstader interview with Bill Klein, p. 3.

[126]*Ibid.*, p. 3, 4.

[127]*Ibid.*, p. 4, 5.

[128]*Ibid.*, p. 5,

[129]*Ibid.*, p. 6.

[130]*Ibid.*

[131]*Ibid.*

[132]Gartner interview, p. 2, 3.

[133]*Ibid.*, p. 6.

[134]Dochstader interview, p. 9.

[135]*Ibid.*, p. 9.

[136]*Ibid.*

[137]*Ibid.*, p. 10.

[138]Gartner interview, p. 4.

[139]Bunnell interview, p. 3.

[140]*Ibid.*, p. 4.

[141]*Ibid.*, p. 14.

[142]*Ibid.*, p. 6.

[143]Mauritz interview, p. 1.

[144]Gartner interview, p. 3.

[146]Noah interview, p. 1.

[147]Nerad interview, p. 3, 6, 12, 13.

[148]Capuzzo interview, p. 3.

[149]Beck interview, p. 3, 7, 20, 21.

[150]Prittie interview, p. 2, 3.

[151]Jim Wilson interview, p. 7, 1.

[152]Tomlinson interview, p. 8.

[153]Wert interview, p. 25.

[154]Donovan interview, p. 10.

[155]Sidrane interview, p. 8.

[156]Buchta interview, p. 5, 6, 9.

[157]Andrews, Jr., p. 2.

[158]Garber interview, p. 1, 2.

[159]Hulton interview, p. 3.

[160]Hattemer interview, p. 1.

[161]Muir interview, p. 4.

[162]Werneke interview, p. 1.

[163]Anderson interview, p. 21.

[164]Kennedy interview, p. 7.

[165]Wise interview, p. 6.

[166]Jim Hart interview with Bill Klein, p. 1.

[167]*Ibid.*, p. 2.

[168]*Ibid.*

[169]*Ibid.*, p. 3.

[170]*Ibid.*, p. 5.

[171]*Ibid.*

[172]*Ibid.*, p. 6.

[173]*Ibid.*, p. 14.

[174]Bert Kennedy interview with Bill Klein, p. 2.

[175]*Ibid.*, p. 6.

[176]*Ibid*,. p. 17.

[177]*Ibid.*

[178]*Ibid.*, p. 4, 5.

[179]*Ibid.*, p. 19.

[180]*Ibid.*, p. 19, 20.

[181]*Ibid.*, p. 7.

[182]*Ibid.*, p. 21.

[183]Chuck Miller quoted in The First Fifty Years, p. 134.

[184]*Ibid.*, p. 136.

[185]John Arozena quoted in The First Fifty Years, p. 95.

[186] Clark, The First Fifty Years, p. 96.

[187]*Ibid.*, p. 102.

[188]Dick Riley interview with Bill Klein, p. 3.

[189]*Ibid.*, p. 6.

[190]*Ibid.*, p. 2.

[191]*Ibid.*, p. 14.

[192]Joe Barnette interview with Bill Klein, p. 8.

[193]*Ibid.*, p. 8, 9, 11, 13.

[194]Larry Kunkle interview with Bill Klein, p. 5.

[195]*Ibid.*, p. 4.

[196]*Ibid.*

[197]*Ibid.*, p. 4, 5.

[198]*Ibid.*, p. 6.

[199]*Ibid.*, p. 6, 7.

[200]*Ibid.*, p. 7, 8.

[201]Scherer interview, p. 5.

[202]*Ibid.*, p. 6.

SECTION SIX
COMPETITION IS THE HERO:
1980-1990

As the pressure sensitive industry began to take form it did not completely lose its rough and tumble nature. Competition in quality and price continued to drive the business. There were acquisitions and attempts to concentrate the industry, and new competitors entered the fray, but the entrepreneurs fought to keep their own identities.

NO SECOND TIME

In the early 1970s, after Litton Industries had forced Jim English to fire his old mentor Seth Wheeler, English made himself a promise that he would never do that again. English was president of Contact Products in 1979, when owner Herb Smith sold the firm to Dixico.

With a wonderful sense of timing, Dixico came to English in December, 1979 with another list of layoffs.

And I said, "No way. I'll save you a lot of money at the top." I knew it was coming...and Jack Pager was looking for a partner at the time, and that's when I made the move to Michigan and got involved in Kalamazoo Label Company.[1]

Today, under an umbrella company called KALGRAFX, there are five companies:

Kalamazoo Label Company, which is roll pressure sensitive. Kalamazoo used to be a lot of offset cut and stack labels. We moved that up to Grand Rapids and started a [sister] company called Sterling Color Process....Then we bought one that makes printing plates, called Michigan Color Plate and Calco Systems.[2]

Jim English, whose work at Fitchburg brought him into contact with many of the leaders of the industry, had nice words for almost all those he has met. But his most quotable quote must be censored a bit for legal purposes;

He's a good guy, but_____was the kind of guy, I think if he told you two words, he told you three lies.[3]

CONSTANT HORROR

So you want to be an entrepreneur? Do it yourself? Make your fortune? Chart your own course? Well, there may be a flip side to this adventure. Bruce Bell had been in the business forms business for seven years before he decided he wanted to be on his own.

> *I didn't know what business I wanted. I didn't want to be in the shoe business or the tavern business....I wanted something that had...lots of market. It had to [have] steady growth during the ups and downs of our economy.*[4]

The Miami of Ohio graduate bought a new five color Webtron 650, hired one person to run the press and another to do office work. He did the selling.

Being in Wisconsin it was natural that Belmark, Inc., would start out with cheese and meat labels. The first year or so Bruce described as "constant horror. Nothing but fear."[5] He remembered his first order as being rather small—about $400—but very exciting. "To be able to do that, invoice, and know that people really paid you for it."[6]

But in his first year he grossed about $70,000. And it was $200,000 the next. "It went pretty well," Bell admitted,

> *Some of the things I enjoyed were finding niches. A lot of the customers that I got then didn't have a need for pressure sensitive labels. They had never used them before I contacted them. Being able to work with them and solve problems—show them how pressure sensitive could be helpful—was fun.*[7]

Bell says service and fast turnaround are the keys to Belmark's success.

RATHER BE LUCKY

Weber Marking Systems had been around for decades making stencils and other direct marking systems. Joe Weber saw that pressure sensitive labels were stealing his lunch. He decided not to fight them but to join them, and started printing pressure sensitive labels in 1980 or 1981. Strangely enough Weber had been coating pressure sensitive materials long before getting into printing labels, probably back in the late 1960s. The company did this because some of the roll stock they were providing to go through Weber label printing machines was pressure sensitive. It was hot melt pressure sensitive adhesive right from the start.

Customers using the Weber electro-mechanical labelers were unsatisifed. They wanted type larger than typewriter size, variable information, and more. Weber realized with a computer and a dot matrix printer, they could have all these extra value features.

We were there before bar codes came along. I'd rather be lucky than smart any time. [Then] bar codes came to the fore. More people wanted bar codes and, as a result, our computer-driven printers really took off. And we decided at that point that we had to get into the software business so we started to write software. We became a software driven company.[8]

System integration is one of the many factors driving Weber Marking Systems.

Computers are getting cheaper and cheaper. The printer is getting cheaper and cheaper. The switch from dot matrix in printing to thermal and now to laser....The switch from nothing to bar codes and the various bar code configurations.

The move on our part to becoming a large software supplier has been a big move. About four years ago [c. 1989] we made a move, we bought applicators and installed printers on those applicators and we became a large supplier of printer applicator devices to print and apply right in line. That's become a very big part of our business in the last three or four years.[9]

Weber essentially produces secondary labels, but they purchased a primary label printer, Tape and Label Engineering, to fill in their line. Tape and Label Engineering is the highly successful creation of Lee Paul, profiled earlier.

Another unusual aspect of Weber Marking Systems is its international operations, dating back to the 1960s. Weber was approached by an English business forms company that wanted to form a joint venture company to take a proprietary stencil which Weber owned, and attach it to their forms.

We started a miniature plant...and we went along with it, and it became very successful. Then we hired this fellow [whose mechanical calculator business had been destroyed by the electronic calculators] and he had all these contacts all over Europe. He set up distributorships for us. We made the products in our English plant and [sold them] all over Europe.[10]

When the English joint venture partner, a publicly-held corporation, became part of a takeover bid in the 1980s, they asked Weber to buy them out. Weber opened a plant in Germany later.

POLITE, BUT STILL C.O.D.

Expansion into the Far East proved more difficult. In the 1970s, Weber and his brother went to Japan to open a company there. The company was formed with the Japanese owning 49 percent of the operation:

We started doing a little bit of business, we hired a couple of people...about a year or two later my brother was running this thing, but [soon] they ended up owing us a lot of money. They're not paying....They're not paying us.[11]

So Weber sent his head of international operations, Bill Darris, to Japan to close the company down. By this time Weber owned 62 percent of it. Four days after Darris left, Weber got a phone call.

Darris said, "They won't let me shut it down."

"What do you mean, they won't let you shut it down? It's our company, we own it. What do you mean they won't let you shut it down?"

"They said the social cost of shutting it down is more expensive and they say we should put the company on C.O.D."[12]

Which is what they did. Today Weber has a company in Japan with two employees whose customers must pay Weber cash on delivery.

But Weber does remember how polite the Japanese had been. He and this senior Japanese businessman were sent off to dinner together in a taxicab. Neither knew a word of the other's language:

At that time everybody smoked all the time. So I pulled out a pack of cigarettes....He took one and I took one. I lit it up for him, I lit mine, we sat there quietly smoking our cigarettes. I put mine out, he put his out. We got out, went into the restaurant, and I found out two weeks later it was the first cigarette the man had smoked in his life![13]

KLEIN ON JAPAN

In 1960 Bill Klein began working for Fasson in Painesville, Ohio. About six years later he moved to Avery Corporation in San Marino, California to become Director of Development for Base Materials Group. While there he began a series of negotiations with Sanyo in Japan. Early in 1970 it was agreed the two companies would create a joint venture to produce base materials for the Asian market. A Japanese national was to head up Fasson's involvement but the Japanese balked. They had been dealing with Klein for two years...

And they'd rather deal with a devil they knew than with a devil they didn't. So I went to Japan as Resident director.[14]

The company was called Sanyo-Fasson and its plant was beautifully situated at the foot of Mount Fujiyama. The plant, built mostly with women laborers, was up and running quickly.

Klein says doing business in Japan is both different and difficult for Americans:

For example, you'd get an appointment with a Japanese businessman and you start out with tea and perhaps an hour of polite questions and answers about your family, the weather, just about everything but business.[15]

The prime Japanese partners found Sanyo-Fasson an ideal dumping ground for executives that did not fit into the Japanese organizations. Klein spent two and a half years there and then returned to Painesville. A Dutch native, familiar with Fasson Europe, took over. The venture's management was later left in the hands of the Japanese partners.

Fasson's Asian operations are now primarily in Australia and Korea, while Avery Label still has a highly successful Japanese operation.

THERE'S US AND THERE'S AVERY

Many of the interviewees would talk about something they had done and sometimes added, "Oh, yes, Avery had done it first." That's because most of the converters thought of Avery as this giant who wasn't really a part of the industry. Often they might acknowledge that Fasson was part of the industry, but the label portion of Avery was not.

Klein says he understands why this feeling existed. "Avery was first and often better and, at times, somewhat arrogant. And since they were never the low price producer they were vulnerable to the smaller, hungrier producers. Avery walked away from some business rather than compete."[16]

Klein, who enjoyed Japan enough to adopt a Japanese son, says he learned just enough of the language to know when he was being conned.

Apparently that helps.

AGAIN, A RECESSION HELPS

The recession of the early 1980s came just a year or two after Stadia Colorado Corporation had moved from its log cabin to a leased building about ten blocks away. Dan Mauritz recalled that the recession made companies more receptive to arguments for printing labels in-house.

> *Everybody was trying to save a buck. So that part of our business really took off and then our printing business took off because we finally hired a regular professional printer and bought a real press, and so our whole business just skyrocketed thanks to that depression.*[17]

Stadia began to grow geographically at about this time, opening an office in Salt Lake City and another in Albuquerque. All the printing, shipping and invoicing came out of Denver. Although they had color facilities, Stadia tended to seek out industrial labels, at least until the mid-1980s.

When thermal transfer printers became available, Stadia was ready:

> *[Earlier] we had been out selling...people had to buy printers, people had to get ink on their hands, something they weren't too accustomed to. Now we had something that was computer driven, needed no plates, it was just phenomenal. We sold hundreds and hundreds of them. And supplies.*[18]

In the late 1980s a plant manager was hired. "We never felt like we made many errors, [but] we made even less with his help."[19]

Stadia considers itself a Rocky Mountain Time Zone operation of about a dozen states.

I don't see any changes in the way we'll be doing business....we'll just be trying to do all the things we feel comfortable doing, looking for little places here and there to tweak it. Try to keep our eyes open to opportunities.[20]

Stadia Colorado must be doing something right. Mauritz is projecting double digit increases even in the slowest-growing aspect of the business—blank EDP forms.

LABELING SYSTEMS AND FLY RODS

Vietnam veteran Lon Deckard has a degree from the University of Kentucky and an MBA from the University of Cincinnati. His goal was to build a business and sell it, but first he worked selling office machines for a year before joining Avery Label. Why office machines?

When I got out of graduate school I discovered that the two industries that were most willing to take people into sales without field experience were office machines and life insurance. I perceived office machines as the lesser of the two evils. In 1974 I went to work for Avery. In 1975 I was their top salesman in the country....They cut my commission rate....I went on my own as a combination rep and broker.[21]

In 1978 Deckard and a partner formed a joint venture label company. The partner was a silk screen printer who wanted to get into flexographic printing. A year later Deckard bought a small lithographic company he had represented. When the partner decided to concentrate on screen printing, Deckard bought him out and combined the roll label company, the lithographic company and the brokerage business under the name Graphic Resources.

He did a lot of business in the toy and appliance industries, but concentrated on building sales in the healthcare industry. Graphic Resources also holds a patent on a multi-ply instant coupon.

The development of the instant coupon and our perfection of the art of printing by the rotary letterpress...were just two absolute technological milestones for us.[22]

Deckard said he believed Graphic Resources is the largest Gallus installation in America.

We started out in an industrial store [in 1979]. In '80 we moved out...bought a building out on the western side of Cincinnati, about 15,000 square feet, and stayed there until '84 when we built our current facility in northern Kentucky, which we have expanded twice.[23]

But expanding his Cincinnati facility was not the only way Graphic Resources was to grow. In 1985 Deckard bought Label Products in Phoenix.

In '87, I believe it was, I purchased Ritter Label in Atlanta, Georgia. Kept that as a separate operation for a while. When the lease ran out the landlord [proved unreasonable] so we just moved it up to the Cold Spring [Kentucky] operation.[24]

In 1988 he purchased a Kentucky label company and "in May of '90 I purchased Layton Label in Heartland, Wisconsin. So now we're operating out of three plants."[25]

In February of 1991 Deckard sold Graphic Resources to Engraph, worked for them for eighteen months and still acts as consultant. From day one, Deckard said, the plan had been to grow the business and sell it.

But Deckard is still in the pressure sensitive business, since he bought Fasson's machine company, which he still operates under the Quadrilleeling Systems name. He also owns a fly rod business (top retail price about $3,000).

Deckard gave customers the credit for building his business:

Particularly the ones that came with us at first when we were nothing. So I give credit to three companies in particular—Kenner Products, a toy company in Cincinnati; a company [then] called Designer Manufacturing, now part of White Consolidated; a third one called the Clorox Company...where we started our instant coupon business.[26]

What was the highlight of Deckard's career? During Desert Storm, "the greatest, most pleasurable thing I ever did in the PS label industry was to produce support our troops stickers. [I made] thirty million of those. Gave them all away—sent them all over the world."[27]

FOR LASERS ONLY

For thirteen years Bob Dillingham worked for others in the Pressure sensitive industry, before starting up his own business, Hi-Tech, in 1986.

In 1973 he went to work as a salesman for Package Products Company, a fast-growing converter of paper and label materials in Charlotte, North Carolina, now a part of Engraph/Sonoco. His territory was South Carolina and Georgia.

> *In addition to labels, they also produced printed laminated bags and pouches, paperboard folding cartons....They wound up eventually printing pressure sensitive by four different processes—by web offset, by flexography, by letterpress, by rotogravure. There was a guy there...by the name of Skip Heintzelman. He's a real innovator...really knows his business and knows exactly what he's doing.*[28]

One problem Dillingham recalled was labels that were not adhering to a client's rebuilt automotive starters and generators. The product was a preprinted label that was imprinted with variable data. Dillingham had sold them a system and it wasn't working. The problem was that the company was adding fish oil to their paint thinner.

After a year and a half in Charlotte, Dillingham was transferred to Cincinnati, staying with Package Products a total of nine years. Then he joined Ted Hattemer's Adhere Products as a salesman and later became sales manager and then, general manager.

> *At the time it was part of a group of seven printing companies owned by Consolidated Foods....I found Adhere to be extremely innovative....It was amazing what type of products they put together...very exclusive prime labels, printing of name plates, sophisticated over-laminates and things like that.*[29]

He left Adhere to work for H.S. Crocker as specialty products sales manager. Then Dillingham replaced the general manager, and a few months later, "they decided to close their plant."[30]

Crocker closed its plant in Blue Ash, Ohio (a Cincinnati suburb) and moved its equipment into three other label plants around the country.

But Dillingham's time at Crocker wasn't wasted. At a sales meeting, Flexcon Company came in and announced the latest and greatest "thing going on in the label industry of today. He holds up this 8 by 11 inch vinyl sheet."[31]

The bar codes were printed with a Xerox laser printer. A computer tells the printer what to print. Dillingham was especially interested in a shelf marking application and was able to sell Kroger their first shelf marking vinyl pressure sensitive sheet that ran through Kroger's Xerox laser printers.

It wasn't an easy product to make because:

After storing for a period of time it had to go through an electronic printer at 100 to 120 sheets per minute, past a 300 degree fusing station and not melt and not bond and not curl and come out the other end with the toner still locked onto the vinyl and not jam the printer.[32]

Dillingham says the problems caused Crocker to back away from the product, which is when Dillingham decided to start his own business in 1986. In the early 1990s he moved to the Fairfield area, north of Cincinnati.

A VERY NARROW NICHE

"One thing that makes us a lot different from the other printers of this product line is that we have web offset and sheet fed offset," said Dillingham.[33]

Most of the other folks have flexographic presses. But we're in a very narrow niche. To our knowledge we're the only company in the country that just produces label sheets for laser printers. Now there are a lot of other people that are in it, of course, and Avery being the biggest, but they also produce roll labels, continuous labels and things of that nature.[34]

Continued improvement of the laser printer will allow continued growth, Dillingham believes:

All you need is a PC and software and...a laser printer and all of a sudden you could just print all sorts of different products. You can print dry gum, label stock, you can print tag, you can print plain paper, you can print both sides.[35]

There are additional computer-driven printing processes, including ion deposition, ink jet and digital sublimation.

I see a proliferation of the end-line electronic printing devices. It makes me feel that in just no time...these web fed presses will have electronic printing heads on them and we'll have no plates, no ink, and they'll be computer driven and it will be a toner technology and we'll print from color toner heads.[36]

THE FORTY-YEAR ITCH

Many of those interviewed for the book worked for others for fifteen to twenty years, then decided to go into business for themselves at about the age of forty. Paul Dunphy of Design Label Manufacturing is a good example. He started production work for an aerospace company right after college, worked there seven years, then spent eleven years as a sales representative for IBM.

> *In 1977 I sought a business to buy for myself because I now had the itch.... I found myself a company called Design Label in Connecticut that was a small company doing several hundred thousand dollars....That was one of my criteria; it had to be small enough that I could cover the down payment. It was a company that I could buy and support [my family] out of earnings.*[37]

Dunphy didn't know anything about the label business except what he learned from the previous owner. But it made money, didn't require a high degree of technical expertise and offered a high potential for growth.

Dunphy credits TLMI with providing a peer group which, he says, "was terrific for my own morale."[38] He says he needed a peer group as it was the thing he missed most in leaving a large corporation.

The company places a great deal of emphasis on service and has chosen to remain essentially a Connecticut company.

> *When you're a small person on the block, service is the most important issue, particularly for packaging type labels. This is not a prime label...service is the ultimate requirement.*[39]

A huge order forced the company to seek additional space and new equipment in the late 1980s. And that's when the recession hit New England.

> *That was tough times in the late 80's, early 90's. We added a great deal to our overhead...fortunately at the same time we had some contractual business that carried us through....We are moving along fine at this point and looking for another threshold to go over.*[40]

Three Dunphy children are in the business. The elder son has been working five or six years. His daughter has special financial skills, and the youngest child has just graduated from college and joined Design Label.

For the first ten years they ran a single shift. Now they're running three![41]

CRAFTSMANSHIP VERSUS TECHNOLOGY?

Ferdinand Rüesch of Gallus believed that the Union Carbide introduction of the first laser engraved ceramic anilox roll really changed the business. It allowed companies like GarDoc to utilize the flexo four color process.[42]

> *Then what really happened [was that] you had more and more [companies] switch from paper to film. So we [Gallus] came out with a rotary screen press.... Bringing rotary screen onto rotary letterpress integrated the technology so that now you could bring UV letterpress and UV rotary screen in combination.... And fortunately Gallus had a design which allowed us to put rotary screen on existing machines [to] retrofit.*[43]

Rüesch further believed this not only changed the mechanical side of the business, but also changed the mechanics themselves:

> *Craftsmanship for many years has compensated for all these inefficiencies....But you know the question was, "Now that certain things started to mature, [could the industry] still afford to do it that old fashioned way? Was anybody paying for the hours and the waste you produced until you had the perfect label?"*[44]

In addition, Rüesch discussed the designs for labels, now "that the big guys discovered the selling power of labels. Now they are coming out with more and more crazy designs, more sophisticated designs...to win the battle for the dollar on the shelf."[45]

Then came another change in printing processes, the UV flexographic press, started in 1985. There was little or no demand for it, however, until supplies improved:

Together with a company called G-Men, a Swedish company which is today on the market under Tosko Nobel Printing Ink....First of all ink became consistent... it's pure and it polymerizes so what you laid out stays down....With finer ceramic anilox rolls, better separation techniques, now...the pre-press technology which became electronic...A computer is the basis of the new tool box....Now [you] work it out [on the computer] and then that result which you worked out remains constant, because the ink is constant. So you can take the job, run it today, you can take it a month later, which wasn't possible previously with flexo.[46]

UPSTAIRS/DOWNSTAIRS

In 1966, with money partly supplied by his school teacher wife, Allan Prittie had started a machine shop. "We fixed golf carts, pots and pans for the local nunnery."[47] He purchased the company for $5,000 down and $19,000 on the *never-never*.

The business was located in Willowdale on the outskirts of Toronto, Ontario.

Just a small industrial [plant], it almost looked like a house. And the chap above us was running a label company...two Mark Andy machines, one two inch and one four inches wide....But the chap above us and I would have a beer on Friday night," and one Friday the label producer [Dan Tomlinson] said he needed a small slitter to augment the two presses. "I didn't have any business and asked him if I could make the machine for him. [It was] the first machine we sold, [and it] sold for a thousand dollars."[48]

That first machine still runs. Arpeco Engineering, Ltd. was successfully launched. Guess who helped Prittie sell the machine? Prittie had built two slitter rewinders, and had four Toronto plants try the rewinder out.

We ended up selling four machines....Then Avery in the United States bought the equipment and then Kimball and Monarch [have] bought a number of them. Still big customers of ours.[49]

In 1973 Arpeco built its first letterpress.

It was a UV curable drying system on the machine, fourteen inches wide....It was a three color machine...and we made some really serious mistakes along the way....Some people still use them...and liked them very much. I mean...Gallus brought their first rotary machine out with UV curing in...1980, I guess. And Ko-Pack a little bit earlier... NilPeter was a little bit later. We were out [with a web] wider than all of them in 1975. If we had been good marketers we'd probably be eight or ten times the size we are today.[50]

Prittie said that in 1977 or 1978 they decided the inking system wasn't good enough. At a 1980 business show in Chicago the press would not work. It was placed at Apollo Label in Chicago where Apollo's Jerry Babick made "a major contribution to solving those problems."[51]

We really almost died through all of that. Basically [we] said, "Why don't you cut back to your original business that was very strong." And it was....But I couldn't give up on those people who trusted me and we stuck it out. And really, it was not until 1989 [that] we made some major breakthroughs.[52]

The major technological changes of the 1980s and early 1990s are staggering. Prittie suggested that the label manufacturer concerned with consistent quality should not "rely on artists running his machine."[53]

AN INTERNATIONAL WAR ROOM

Since Canada has a small population, Arpeco has ninety percent of its output exported to Japan, Australia, New Zealand, the United States, the United Kingdom, Ireland, Scandinavia, Italy, Germany, Switzerland, Austria, and is now selling in South America.

We are communicating with our machines by computer link to do the trouble shooting. We have a little war room here—the whole concept is to have your senior designers [available]....We can't be in [all] corners of the earth at the same time. So if we have a problem in the field...we'll get our senior people in one room and solve the problem by phone...any time day or night.[54]

WHY 7 1/4 INCHES?

Allen Prittie flies to Europe almost monthly...

And when we'd go to Europe the Avery unwind shafts were 7 1/4 inches in diameter. And the Europeans thought this was madness....I never talked to Stan Avery so I don't know if this is true or not [but] I was told by some old Avery guys that the reason is...Stan Avery originally wound up his labels on a paint can.[55]

BARELY ENOUGH TO BUY LUNCH

Not all entrepreneurs stayed with the label business, but just couldn't find a better industry to participate in, one way or another. After nineteen years in the business, Del Styers and his two sons sold Emblem Tape and Label in 1983. Shortly thereafter, son Paul Styers started Styers Equipment, which sells used printing equipment. Why did the three decide to sell the label company? Paul explained:

> *The average entrepreneur found that he's building his business...maybe in good leaps like twenty, thirty, fifty percent a year. And it's a billing business. You've got a bunch of invoices out there....So your receivables go up, which requires capital. Doing more business, your inventories go up. You're making the bucks, so you have to pay your taxes....The average entrepreneur in this business, he's...going great, this is wonderful—a couple million dollar business and I just got barely enough to buy lunch.... You are investing everything in the business. You bought the new press. And the new press was no longer $14,000...it's $50,000. Anyway, we built the business and we tried to operate it as frugally as we could. We saw an opportunity to sell it and put a few bucks in our pants instead of having them all tied up in the business.*[56]

Paul Styers says government seemed to interfere more and more. IRS audits. Employee complaints.

> *I ended up doing everything but what I really thought was important, which was getting that customer's order out the door. We had one employee file a discrimination suit that took a lot of the wind out of me. The court threw it out in five minutes [but] it took five years to get to court...so I kind of suffered that kind of burnout.*[57]

The three Styers sold the business on January 14 and Paul incorporated his used PSA converting equipment business in February. He is currently celebrating ten years of successful operation and is bullish on the pressure sensitive industry "Tag and Label [is] the only facet in graphic arts that is experiencing growth today."[58]

THRASHING AROUND, BARELY SURVIVING

About sixty percent of the product McCourt Label produces is fan folded for computer use. It was the purchase of a wide web press in 1980 (plus a second a few months later) that jump-started McCourt Label.

The company had been around for decades and was inherited by John Egbert's father in 1960. In the mid-1970s his father sold the company to John and his two brothers, both of whom subsequently left the company. John had worked for his father, an absentee owner, as sales manager for a few years before the sale. The company was not in good shape, but had managed to straighten itself out in the later 1960s when

> *pressure sensitive became commercial. I think at this point he [John's father] needed somebody in sales and that's what I did for a living. I went to work for my dad in '70 as a sales manager. When I came, the company was just under a million dollars with 84 employees, and today it'll be something between twelve and thirteen million with 99 [employees].*[59]

In 1970 most of McCourt's sales, over ninety percent, were to pharmacies, and that was the era when the big chains were squeezing out or acquiring the independents.

> *What we were selling to an independent drug store was pressure sensitive labels for twenty dollars a thousand. The chains wanted to buy [a thousand] for two and a half [dollars]. I said very quickly, "Well, if we're going to stay in this business...and survive, we need to sell something else to someone else."*[60]

They searched, and survived, but barely.

> *[We] kind of thrashed around until in the late 70's we were doing about three million dollars, I think, in 1979, and in 1980 we bought our first wide web press and entered the computer market.*[61]

The second press came later in the year and by early 1981 both were running two shifts. In July a third shift was added and another press ordered. When it arrived they had it up and running within 48 hours.

TURNING DOWN FOUR MILLION DOLLARS

McCourt Label won a government printing office contract for the postal transfer label. "It's the yellow label that moves your mail from your old address to your new address."[62] This was in 1985 and required two truckloads of pressure sensitive paper each month. They were losing their shirts on it...

because I don't think you get those orders unless you bid them too low. But it taught us a bunch of things and put us in a position to buy paper a little better, kind of learned quick enough to get out of it. I think we had the contract two years....We had run the company up...to about ten million plus in sales...the government business alone was four million....[Not bidding] dropped us to about seven million.[63]

But the government business did provide cash flow and taught the company how to respond quickly to a customer's needs. In July, 1987, John became president and has more than made up the loss of the giant contract. Egbert says of the technological changes in recent years:

The development of water-based inks and the quality you can print with it, the development of water emulsion adhesives have made our industry more environmentally...friendly. In my opinion, the industry must be environmentally aware or sooner or later we're going to be eaten up. You can't keep making stuff out of [oil] and stuff like that and continue with today's world.[64]

On changes in technology:

For example, if you stop and start the Webtron that we bought in 1982, you're going to chew up a couple hundred feet of paper putting the press back in register. If you stop our new Webtron and start it back up, it will start in register. You just have a few feet to ink it up...and I mean a few feet—ten to fifteen—and the press will be back running good product.[65]

One way to lower costs and increase profit is to lower waste. In 1948 the scrap rate was about twelve percent. Today it is less than nine.

Egbert had some nice things to say about one of his suppliers, S.D. Warren, and Warren's influence on quality improvement.

Everything they do, their quality is just a little bit better, and it didn't used to be that way. It used to be they were a lot better. And now everybody is beginning to play catch up. In my opinion they're the ones that have taken the industry and raised its level substantially.[66]

33 EMPLOYEES, 300 YEARS' EXPERIENCE

Jim Sauer, of Retterbush and Sauer Label Corporation, started working for National Label in 1965, fresh out of high school. His job was "the lowest job, the nastiest job"[67] National offered, cleaning the rubber plates. After a while he was shown how to run a press. National built some of their own printing equipment and began selling these products under the name of International Machine Products. Sauer moved to that division and installed the presses. A year or so later he became a very knowledgeable label salesman and found his niche.

Sauer says, because of solvent-based inks and coatings, they were averaging a fire a week in the late 1960s and, although the fires were seldom hard to put out, it would often take five or six hours to clean the presses after using a fire extinguisher.

In 1974 he took a sales job at Adhere Products in Cincinnati and worked there four years. In 1978 Sauer and fellow Adhere salesman Don Retterbush decided to start their own business. In 1978 they pooled their cash, "what little there was," said Sauer,[68] and started Retterbush and Sauer Label with one used three color press.

At first Retterbush sold and Sauer ran the press because of his mechanical experience, but soon an employee was added and Sauer returned to selling.

> *We were off and running right off the bat. When we left Adhere Products we both left on [the] uppity-up. We were not under any contract situations, we did not set up things with customers before we left. We tried to do it as first rate as we could. I've always been kind of proud of that fact.*[69]

> *We both left sales fields where we had known people for a long time so we felt we could gain some business rather quickly, and we felt people would be loyal to us personally...and they did that rather quickly, and so that was a big help for us.*[70]

Growth was swift. In fact, Sauer says, "We were growing a little faster than we wanted to."[71] He had promised his wife when they started the business that he would not turn into a workaholic and he has not, noting with pride that he has worked less than ten Saturdays in fourteen years. But the rapid expansion required more equipment and more people to run the business.

I was fortunate at being in this business all my life here in this town [Cincinnati]. I knew where there were a lot of good people. I've always believed a company is really its people, not machines. I always felt if you surround yourself with really good people, you can't lose....We're proud of the fact that in our fourteenth year of business we have 33 employees [with] over 300 years of experience, which I think is kinda neat.[72]

WATCH YOUR ACCOUNTS RECEIVABLE

Still, in the early days of Retterbush and Sauer a Cincinnati pharmaceutical firm came to them and wanted to place all its label business with the firm.

We had called on them for a while...and could hardly get past the front door. We weren't smart enough to figure out why they all of a sudden [changed their minds]....We started doing a lot of work for them and got into the tune of them owing us about $35,000. Then we found out they don't pay...all they did was use us until they could use us up. This was a really tough time for us.[73]

The money was finally paid (just before the pharmaceutical company went bankrupt) but the young firm had to get some credit extensions to keep afloat.

Retterbush and Sauer has not jumped to buy rotary letterpress or rotary screen although the company continues to evaluate the potential. The company currently operates nine presses and has never borrowed to buy one. When it comes to a million dollar press, Sauer said, "I can't sit here and tell you that I could just run out and pay cash for one of those babies."[74]

We try to pick our spot, we're probably a super, super aggressive sales group....We just try to pick our spot. We feel we have the right kind of equipment to pick the kind of spot that we want to pick.[75]

A NEW GIANT

Engraph, one of today's leading PSARL producers, wasn't much of a player until the mid-1980s. In 1981 Engraph's chairman, T.J. Norman, decided to retire and the board chose Leo Benatar to become CEO.

Benatar had already spent a career (28 years) at Mead Corporation, finally serving as head of the giant paper maker's packaging division from

1972 to 1981. Prior to that he helped Mead open up an international operation in Europe. Engraph's only pressure sensitive connection in 1981 when Benatar took over was a small plant with rather "antiquated equipment."[76]

> *I was intrigued with the pressure sensitive labels. At that time stylized containers were coming in and I just saw the opportunity to have a position. [At Mead] I used to spend a lot of time in the supermarkets....As I saw the industry changing...I just saw that labels were going to be a more important factor than they had been in the past.*[77]

Engraph's lone plant was doing about three and a half to four million at the time. Although they considered acquisitions, the company decided to grow internally at first. It was 1987, Benatar recalled, when they acquired SGI Memphis. SGI was about a fourteen to fifteen million dollar business at the time:

> *And it was not a well-known pressure sensitive label company, but we saw they had some unique features. They had some good equipment, [they were] very, very customer oriented, and about forty percent of their shipments were made within one day of the time they received the order. It was unique for the industry.*[78]

The following year Benatar talked with Alfred Industries concerning Alfred's Patton operation. Patton was doing 30 to 32 million and Engraph acquired them in 1989 "and this really gave us a big position because of their reputation. This really thrust us into the pressure sensitive label area in a meaningful way."[79]

Patton had two operations, one in New Jersey, the other in California.

> *And then I think it was in 1990 or 1991 we acquired Graphic Resources. They had three facilities. One in Cold Spring, Kentucky, one in Phoenix, Arizona and one in Hartland, Wisconsin. Their business was...about 25 million....Then last year [c. 1992] we acquired Polaris, which is one of the leading [PSA label] producers for the cosmetic industry. I think they were doing about twelve or thirteen million....In the interim, we set up a joint venture in Puerto Rico...that does three or four million in labels for the healthcare/pharmaceutical industry.*[80]

Engraph is very much interested in rotary screen printing, focusing on healthcare, coupons and other consumer product packaging. Another trend, said Benatar, is the extended text label. Partly due to government regulations, the labels are required to add more and more information. In some cases an extended text label is replacing a carton or a separate carton insert.

The recycling of packaging materials is a serious problem in general, and PSA label release liner recycling is a particular problem.

Somehow we've got to be able to recycle that material [the silicone liner] or go to a liner-less face stock. We do have a challenge out there and we've got to come up with some solutions....I think we can do it.[81]

I COULD TELL IF I HAD A GOOD YEAR BY THE NUMBER OF CARS IN THE GARAGE

Practically everyone who has ever attended a TLMI meeting in the past 25 years knows Dick Schwartz of Aladdin Label, Inc.

A somewhat high expense account for a tag and label meeting led to his leaving Ford Label Company, even though its owner had been like a father to Schwartz during his five years at Ford.

Schwartz incorporated Aladdin Label in 1972 and at first did almost everything himself, along with Karon Star, who became the first employee a week after the company opened its doors. (She is still with Aladdin.) A leased Webtron 650 was his first piece of equipment. The money needed for the start-up came in part from a company he established during his days at Ford Label, because:

I couldn't even get them [Ford] to paste up a straight line of type [so] for my roll pressure sensitive customers I had to form a side operation called Schwartz and Associates....So I ended up doing it myself at home.[82]

Schwartz's original plan was to have ten customers doing a million dollars each.

Okay. It didn't work out like that, but we ended up getting them anyway. But we had mostly big companies for our customers. We were heavy, and still are, in the food industry. Namely the cheese industry, mostly in what were called Hobart style labels.[83]

There was really no take off [year]. It was gradual because I'd learned enough in five years of working for someone else that I didn't want the tail wagging the dog. And we grew at a steady pace....I remember the first month I had $20,000 [in sales]. [We had] a bottle of champagne. I remember the first month we had $100,000 and I remember the first million dollar month.[84]

Our numbers have been outstanding. The bank looks at our managerial statements and they die. I don't think there's anybody better consistently over the long haul. There's nobody better in the business than Aladdin Label.[85]

Schwartz says he used to report to Gene Singer that he could tell if he had had a good year "by the number of cars in the garage."[86]

PASSING ON THE TORCH

In 1983 Schwartz selected two of his senior officers and allowed them to take control of the company. He says they poked along for about five years "with things as usual, and in the last five we've made many technological innovations that have put us right on the cutting edge."[87]

Early on, Aladdin was a profit sharing company, a somewhat unusual practice in the industry. When his son admonished him after Schwartz had spoken of this practice at a TLMI meeting in Puerto Rico, Schwartz told his son, "It's easy to tell them, kid, because nobody else is going to do it."[88]

THE TROUBLE MAKER

Dick Schwartz characterizes himself as a "trouble maker and then a radical"[89] concerning his participation in TLMI.

It was real clique-y and there were a couple of people in power. [Gerry Osman] got me on the board through hook or crook and I was not a practical choice going in. And I've outlasted everybody and I have personally changed the organization through my personality....I started out...being a rabble-rouser...Skip Heintzelman from Package Products is one of my...supporters. He's watched me over the years and he couldn't believe the change from being Mr. Troublemaker to the king maker.[90]

It took a personality like me. I'm aggressive, I'm flamboyant, and I make friends easily and take friendship seriously. And because of that all the new up and coming guys...at TLMI gravitated to me....We were do-ers and we said, "Wait a minute—this is in our best interest to get our act together." I was there the day Avery's Bob Fletcher walked out. Avery dominated TLMI and they did a great deal for the industry but they started looking over their shoulder and said they are controlling the industry, and they took a hike.[91]

First Fasson and then Avery have since returned to TLMI. Dick Schwartz played the same game himself, walking out of a board meeting concerning TLMI's *Glossary* publication.

IT WOULD TAKE AN EXPERT AMONG EXPERTS

Schwartz regarded the changes in the inks as having had "the major impact on the industry."[92] Aladdin remains solely in flexographic printing, but Schwartz pointed out that few of his associates in TLMI remain 100 percent flexo in spite of the fact that today flexographic printing quality is so high it would take an expert among experts to spot the differences.

He also predicts large gains in flexography in Europe, but says they have a long way to go because "here in America, guys like you [Bill Klein] and I may not agree about everything, but we are a homogeneous group. They are not. They have cultures that hate each other."[93]

UNKNOWN INTERNATIONAL GIANT

Patrick Masterson has been with Topflight Corporation since 1956. He began selling self-wound printed pressure sensitive tape, using Heidelberg letterpresses to make Dad's Old Fashioned Root Beer advertising. From there Topflight moved into PSARL and the industrial field.

My inclination was always towards the industrial side of the business and when we came out with, for instance, the flexible vinyl with the one mil polyester or two mil polyester laminate, then I could go into industry and start to replace metal name plates....A good example is [hand held tools like] Black and Decker....UL approved materials. We went head-to-head [with Avery] on [the fire extinguisher industry] for a long time...and Avery and Topflight pretty well split that.[94]

Topflight began diversifying as early as 1960, when "we formed a sister company called Adhesives Research."[95] It started in a garage behind the Topflight plant in York, Pennsylvania, and is now a good-sized PSA coating operation in Glenrock, Pennsylvania.

Masterson pointed out that Topflight has a network of international plants:

> *In Columbia, South America, we have Topassa, which is a Topflight type company, narrow web label printer, and there is a coater much the same as Adhesives Research. The same two types of business exist in Venezuela, where Telvanka is the Topflight of Venezuela....Taracotta is an adhesive company there. We have Pacific Label in Australia...both types of businesses....We have Topflight, Italy...Tovenka, Switzerland... Topflight, Sweden....we have a licensing arrangement with Johnson & Johnson in India....The international total would equal our state side [total].*[96]

NO LONGER STAND-ALONE TECHNOLOGY

Masterson noted that flexography "is no longer the dominant stand-alone method of producing labels...[Topflight uses] Kammann Presses that are strictly roll to roll screen...a Gallus combo [to] benefit from technical information from other areas."[97]

"When you look at multi-technology presses where you have flexo, rotary letter press and rotary screen on the same press...[flexography] is still dominant, but it doesn't stand alone."[98]

A PRITTIE PRESS

Allan Prittie at Arpeco Engineering, Ltd. "made the very first acceptable rotary letterpress in 1980. It's a very unique system"[99] Prittie points out that the date of the original press was 1975, but it required lots of work before it was successful. Bud Gray of MPI also points out that the converters were responsible for the pioneering work leading to the rotary letterpress. Suppliers ignored it. "The DuPont representative laughed out loud."[100]

Due to converter requests, the plates got better and new inks were formulated. Gray notes the biggest problem was curing the inks at high speeds. If the curing lamps were hot enough, they would activate hot melt adhesives while the web was going through the press, or, if not, the ink wouldn't cure. But, they finally got it right.

> *[Label-Aire] did make a darn good labeling head. The P84 did a good job in putting labels on and helped us move a lot of labels....in the early eighties this was a tremendous impetus to the industry, because as they began to develop their expertise and their abilities to make faster and more accurate labeling equipment, our ability [make quality labels] continued to grow, until finally in the late eighties everything came into position. We could not only print a picture with a woman with her hair blowing in the wind running across a field of daisies with a collie at her side and have it look like a lithographed-produced job, we could also apply the labels on front and back at 250 a minute....It wasn't until that point that the packaging industry accepted us as full-fledge members.*[101]

COATERS

When emulsion adhesives were first introduced they were formulated to be run on the same coaters used for solvent adhesives. Coating methods, drying process, surface tension defects, wetting agents—all were problems to be solved. In the 1980s new coaters designed specifically for emulsions were introduced. Emulsion adhesives are generally regarded as better for die cutting, and are used with high speed applications such as EDP labels. Technicote, which started business in 1980, has used emulsion technology exclusively.[102]

THE LASER ANILOX ROLL

As with many other technological breakthroughs in this industry, the laser engraved anilox roll came from two different sources at approximately the same time, about 1981. A man named Maurice Buckley in Cheshire, England, and a company in Wisconsin were independent creators of the new and much improved roll. Buckley sold his worldwide rights to Union Carbide, bringing another giant corporation into the PSA roll label industry as an important supplier.

Now, in the area the size of a postage stamp, there are 160,000 cells, "each one identical in shape, size and dimension,"[103] delivering the ink to the printing plate. This control has enabled flexographic printing to deliver quality printing almost—some say within one or two percent—the quality of letterpress or offset.

A SLITTER FOR REYKJAVIK

After Didde Glazer fired its Webtron regional sales managers in 1987, Leon Beaudoin thought about what he could do. Since all of his experience was in the equipment end of the business, he opened a used-equipment business.

> *All my contacts in my thirty years were in the machinery and the buying end of the machinery. And my contacts and peers were in the same field....I get on the phone in the morning here and I can talk to someone in Los Angeles and Waco, Texas, and Charlotte, North Carolina, and Atlanta, and they know where machinery is for sale that nobody knows about. My [sales] bulletin goes all over the world and we have people coming in from England, [from everywhere]. We sold into France, we're doing a lot of selling into South America and Mexico....But we get calls from [Reykjavik] Iceland. This has become in two and a half years, worldwide.*[104]

Beaudoin says it would take at least ten employees to do two million in label sales. "We're going on two million this year, and I've got four people, and some of them are part time."[105]

One employee is his younger son, who, while Beaudoin's interview was being taped, was flying to New Orleans to show equipment to some prospective buyers from England.

FROM ZERO TO $65,000,000 IN TEN YEARS

Hopping a plane out on Sunday night and flying home Friday for a year and a half for IT&T made John Patel an unhappy camper. A fellow member of the National Guard suggested MACtac might have an opening, so Patel joined them in 1970 as credit administrator for claims and credits for the graphic arts division.

And if you want to learn an industry pretty quick that's where you can learn it, because you find out in six months everything that doesn't work.[106]

By 1972 he was product manager of MACtac's sheet line and two years later, in 1974, left to join Chemtrol, again as sheet product manager. A fellow MACtac employee, Dirk Desanzo, also went to Chemtrol at the same time.

Both men left Chemtrol for Fasson, Patel in 1976, Desanzo in 1978. Desanzo then then went into business on his own, opening Technicote in 1980. He purchased a used coater from Monarch Marking (Desanzo says thirty inches, Patel says 26 inches wide), installing it in a plant near Akron, Ohio. It was up and running in three months. Why the extremely short time? "That's because we were starving."[107]

Desanzo invited Patel to see the operation. Patel says Desanzo told him, "You know, I can make a product but I don't know how to sell it."[108]

About four months later, Patel joined Technicote:

So I left Fasson and became a legal partner in Technicote. Based on what I'd seen in the industry, it looked like there were some really nice niches...specifically in the data processing area. [Other companies] were working in [the field] but lead times were very long. That 26 inch coater had a lot of versatility... pump out some nice custom volume in quick turnaround.[109]

MONARCH FOR A FRIEND

Shortly after selling his interest in Chemtrol, Bruce Motter had become involved in Technicote. There was very low capitalization. They had bought the used coater for just $35,000.

It was a hot melt coater that was converted....For years in the industry the solvent coating in itself was sort of capital intensive. You couldn't get into the solvent coating business for pressure sensitive without a significant capital investment in machinery. After the introduction and success of acrylic emulsion and hot melt you could then get into the adhesive coating business with a relatively minor investment.[110]

They had a coater, but needed more equipment:

We needed all of our capital for a slitter, to split the roll down to a usable size for our customer. Monarch also had a slitter down in—of all places—Australia....I'm going to guess that we paid maybe $25,000 for the piece of equipment, plus an additional $3,000 to air freight it back to us. We <u>contracted</u> to pay $3,000.[111]

The air freight bill wasn't $3,000, it was $30,000, which Monarch had to eat. But Technicote had a slitter.

We were working out of one office....We didn't even have any heat for the office, so I remember going to K-Mart and buying two of those little plug-in heaters. And we, literally in January, February and...March, had to wear coats inside the office during business hours. It was so cold. I remember we did about $200,000 in that first quarter of business.[112]

The original sales plan was to be at about two million dollars in five years, but Technicote reached that goal about the end of the second year.

A PLANT AND A COATER

The company continued to grow dramatically, adding a plant a year for about three years.[113] George Corbett, another Fasson executive who would join Technicote in 1985, tells how the company acquired its third plant. (Its second was in Terre Haute, Indiana.) The company needed another coater. Who you gonna call? Sure enough, Monarch Marking had one.

Why Monarch? Let's look back. In the middle 1950s Monarch decided to produce their own pressure sensitive base materials so they bought Contact Products in California and moved the equipment to Miamisburg, Ohio. (Contact Products was the firm created just after World War II by three ex-Avery employees, including Stan Avery's college friend, Eliot "Digges" Graves).

As many large corporations have discovered, running a profitable PSA base materials operation is not easy. Monarch was willing to sell the ex-Contact Products operation, now called Presto Adhesives. Technicote...

sent a man down [to Miamisburg, Ohio] to negotiate the purchase of the coating line. They told him, "We don't want you to buy the coating line, we want you to buy the building and *the coating line." [The Technicote man] wasn't one of the partners, so he wasn't authorized to buy any buildings. They basically told him, "You don't understand. The price of the building and the coating line is going to be lower than just the cost of the line. We really want you to have the building."*[114]

So, from a tiny building with all of its management working out of a few dozen square feet of unheated office in Cuyahoga Falls, Ohio, they had space, a lab and "all the things you need to run a pressure sensitive plant."[115] This plant, in Miamisburg near Dayton, Ohio, is now the company's headquarters.

About 1985 Technicote decided it needed a California office. George Corbett, Dirk Desanzo's boss while at Fasson, got the call.

So I came on board to go out and put in a coating plant on the west coast. [They needed] somebody who they knew well and could trust as a business person and manage it, and not go crying wolf and having all kinds of crazy problems.[116]

The plant was located in Corona, California, just south of the Ontario airport and not very far south of Fasson's plant in Cucamonga.

One of the things that Technicote does differently than the other guys is we have some narrow coating machinery. And we'll run small custom coat orders on narrow web so we can do small quantities and we turn around pretty fast. At that point the Cucamonga [Avery plant] had been changed over to do all sheets. So we had the only roll label coating line on the west coast for quite a while.[117]

The California plant has been forced to move to larger quarters. When they moved the coater and two slitters, they were up and running in one week. All the plants are run as separate cost centers, totally responsive regionally. "The hard part for us was managing multi-plant locations....Then we linked the company all together by computer. We're on line 24 hours and you can get the information that you want."[118]

The plan was for Corbett to stay on the west coast until one of the partners (Corbett, Desanzo or John Patel) decided to retire. In 1988 Patel decided eighteen years in the industry was enough, so he retired to Sanibel, Florida. He says:

I could have retired very comfortably in Ohio, but [here where] they start selling land per square foot as opposed to acres in Ohio, it became pretty clear very fast that it was going to be real expensive to live down here in the style I would like to. So I ended up going into a completely different business (screen printing sportswear).[119]

However, Patel is now back with Technicote. Since May 1993 he has been vice president of marketing and sales of Technicote Westfield, Incorporated, a subsidiary of Technicote. (Technicote acquired Westfield to get into the sheet pressure sensitive business.)

Corbett and Desanzo are still in the business, but have an office in Naples, Florida, which they use about half the time. There is now a fifth Technicote plant in Texas.

MONARCH

R.J. Klein, vice president of Advanced Development for Monarch Marking Systems, says there were essentially three players in the gum-applied label business: Soabar, which sought the industrial business, Dennison, which did the fancier, more decorative labels, and Monarch, which dominated in the retail area. Monarch, acquired about twenty years ago by Pitney-Bowes, is a very old company that started in the ticket (tag) business in 1890. Industrial engineer Frederick Kohnle had created a mechanical pin-ticketing machine to help retailers mark their merchandise.

About 1915 Monarch was marketing a gum label dispenser. It is possible that Monarch may have been dispensing pressure sensitive self-wound strip labels as early as 1938. They certainly were an important early user of PSA materials.

Gary McKinney of Cincinnati Precision Plate Company used to work at Monarch Marking. He pointed out that the company has two divisions, the price marking and the prime labeling divisions. Although "they moved it [prime labeling] out of the building for a while, now they have moved it back in."[120] He noted that in 1973 Monarch bought the first photopolymer plate making system.

The photopolymer plate single handedly forced flexo and the pressure sensitive industry to the level it is at today. It forced the ink manufacturers, anilox manufacturers, and the press manufacturers to catch up. And they finally did....The materials had to stay up with the rest of the technology. I think the materials out there today are pretty darn good.[121]

R.J. Klein sums it up:

Monarch Marking Systems has done rather well. They manufacture in the United States, Canada, Mexico, England, Australia, Hong Kong and Singapore. Their bar coding systems are marketed in eighty countries with an installed base of more than 100,000 bar code printing systems.[122]

DEALING IN A TIME-URGENT MARKET

Technicote produced nothing but roll labels in both papers and films. Between 1981 and 1991 it grew to be a $65,000,000 company. Technicote did a very fine job on piggybacks.[123]

It became the prime piggyback company producer in the country, taking most of that business away from Fasson. They also did a very fine job in pattern adhesive coating where they could do small runs much less expensively [with] faster turnaround....Whether you're an adhesive coater or whether you're a printer, [the more] you're dealing in a time-urgent market.[124]

Actually, says Desanzo, "We compete with ourselves.... If we do what we're supposed to do, then we'll keep our customers.[125]

ENTREPRENEURS AND PROFESSIONAL MANAGERS

Motter believes most successful companies in the PSA industry are the entrepreneurial ones:

It takes the mentality, the continual day-to-day risk decision making [mentality]. Many of these entrepreneurs reached a certain size [where] they then were unable to manage their business....They deliberately recognized that and tried to make that change...and brought in professional managers who were not brought up with the industry. Many times that was catastrophic.[126]

Basically Motter is correct. But outsiders <u>were</u> able to make a difference in a number of companies. For example, Russ Smith with Avery, a whole new staff for Timemed, etc. The difference was leadership, the ability of the entrepreneur to let go, and integrating new ideas and people into the company.

SUPER BOWL TRIP UP

After six years with York Tape and Label, Carl Williams moved to Dennison. In 1980 Dennison was a big company, about 700 million, but only a middle-sized converter with about fifteen million in direct converting of self adhesive products. Williams was a "Labeling Systems Development Specialist,"[127] working with the engineers on systems installation for customers.

Now Dennison had a partnership with a company named Syncro-Motion....a very small company that started making pressure sensitive label applicating equipment back in about 1966....and they [Dennison] had the exclusive marketing rights on the equipment. Then we would cut an order to Syncro-Motion and would get the label business, and of course we would mark their equipment up and basically that was it....Systems selling.[128]

Williams tells of his introduction to Dennison and the systems concept:

I started with them in January, 1980, and they used to have what they called the Super Bowl Trip....So mostly everybody was gone including my boss and we had a particularly irate customer that needed to see someone immediately because they had a labeling system that didn't work....I was dealing with the vice president who was the owner's son. [I had] to go up and make this work.[129]

Williams successfully solved the problem, and became an advocate of systems selling.

Dennison was trying to get into promotional labeling and Williams soon learned that "you just basically make suggestions at the right time and try and get people to buy into a concept."[130] The systems approach required that the Dennison representative put himself into the shoes of the buyer. "If someone is buying ten million labels a years, you should go back and...find out what they're doing with the labels."[131]

BAR CODES

In the early days of bar codes few converters were able to print bar codes that scanned consistently enough to meet customers' requirements.

> *When the UPC symbols [came out], the tolerances were plus or minus 9,000th of an inch. Well, I plucked two hairs out of my head and held those to the light and said, "This is a market that no one in the flexo industry will ever be able to compete in" ....Then [in the early 80's] we began to find out that, "Yes we could do it."*[132]

MPI's Bud Gray was right, and, among the early bar code printers, Dennison and Xerox were partners in a company called Delfax. They came up with a very economical way to produce serialized bar code labels.

> *So anyway we found a lot of pockets of major business...like UPS....Because they had the capability to take the rollable labels and at the unwind station have the ion deposition printing station that will print a bar code and make every impression different. Then go through the regular flexo press...do what you wanted to do.*[133]

"We were selling systems and solutions where our competitors were selling labeling equipment."[134] Williams says this was most true in 1986, 1987, when K Mart came to Dennison with a problem: how to get the inventory into the stores, but keep a low inventory in the warehouses. K Mart was using about a million labels a day, running on Dennison printers, with Dennison labeling stock. Williams knew they had to move to bar codes to accomplish this.

> *We were just taking one technology, thermal transfer printing, and replacing another technology that they had purchased from Dennison. So we were losing a lot on one end, but gaining a lot on another end. And these systems are $75,000 a crack. They buy them by the two dozens. A lot of labels, a lot of software, the account service people being there around the clock, and that's what it takes.*[135]

In 1990 Dennison was bought out by Avery. Williams said, "I was to know the value again of trying to work things into a team concept and never over-playing things...to get that credibility with the people that really makes any organization. But it worked out well."[136]

FROM PRESIDENT TO LORD

Alan Percival had joined Fitchburg Coated Products in 1968 as a chemist. His marketing experience in England helped him move into marketing and sales where he became a vice president. The last two of his eleven years at Fitchburg he served as president of Litton's pressure sensitive base materials company.

When he joined Fitchburg sales were perhaps seven million dollars, half of that in pressure sensitive materials. Percival says the change he remembered best at Fitchburg was the unrelenting pressure for lower pricing.

> *When I started producing pressure sensitive [material, it] was only a novelty, but by the time I left, it had become very much a commodity. When people go out to sell Kromekote, everybody had to [offer the same] and you ended up having to compete with price. Anybody was prepared to sell....at a loss. Over the years you saw a number of companies go out of business.*[137]

Shortly before Percival joined Fitchburg, the company moved its operations to Moosic, Pennsylvania, when Simon Adhesives was merged into it. After two years as president Percival left to start Lord Label East in Scranton, Pennsylvania, a few miles up the freeway from Moosic.

> *I don't know how much you know about Lord Label, but it was founded in 1965 by Lester Ordman and Maynard Louis. Maynard had some label experience working for Dennison and Lester worked for Wide Web Flexo Printers.... They had a unique method of growing, by establishing resource plants in partnership with someone in the industry. I started the eastern one in partnership with them....It was good for me because I had to put up very little money. And it was quite a successful venture.*[138]

Lord Label was sold in April of 1989 to an English conglomerate named Porter Chadburn. Percival believes Porter Chadburn is the largest supplier to the dairy industry and the water industry. It operates a dozen plants. "MPI is probably the [competitor] that is closest to Lord Label in character."[139]

Percival remains with Porter Chadburn and he and his key production man, Dave Lecates, are continuing to build their business. They are heavily involved in thermal transfer printing. Both thermal transfer and film labeling are among the fastest growing trends in the PSARL industry.

A BEAUTIFUL MARRIAGE

Business forms of one sort or another have been big business in the United States since the late 1950s. They have always attracted aggressive and innovative sales forces. Standard Register of Dayton, Ohio, was among the first to realize that pressure sensitive could help them create a value-added form.

They probably started adding some pressure sensitive material to their forms in 1965 or so. These were farmed out to converters such as Ever Ready Label and others. There wasn't much in the way of volume for a number of years. Standard didn't begin a serious investigation until 1974. In April, 1975 they began negotiations with Universal Tape and Label in Cincinnati to purchase their assets, also buying a prototype hot melt coater, rewinders, folders and so forth, including three Mark Andy presses.

By 1978 they were operating at full capacity and in 1980 purchased the assets and experience of Lambouy Unique in Terre Haute, Indiana. The company was shipping seven and a half million dollars annually and, after the purchase, five and half walked out![140]

To solve the problem, Standard chose their highly successful salesman/product manager, Lance Hailey.

Hailey had started work at eleven years old and did everything including attending—but never graduating from—five colleges or universities. He answered a newspaper ad and sold for Standard Register in the Boston area for seven years.

Hailey wrote $1.2 million in sales in 1967-1971, earning a substantial $60,000 or more a year (Standard doesn't care if salesmen out-earn the president). When Standard bought the Terre Haute plant with few customers, Hailey took over and by the early 1980s, was growing at thirty to forty percent a year.

In the mid 1980s Hailey was beginning to merge forms and PSA with inventory control/picking tickets. For the fancier forms he used outside suppliers including MPI. It was, Hailey said, "a beautiful marriage."[141]

Other forms people didn't get into it to the degree that we did, except maybe Uarco. They had a very knowledgeable guy, Bill White; they did great things like we were doing, got into it earlier than we did. Moore was there, but only in stock labels.[142]

By the early 1990s Standard Register was doing $100 million, with about a quarter of this collated into business forms.

Hailey saw thermal and laser growing rapidly together but impact printing on its way out. He credited Roger Clark, an "engineer type, my Man Friday—he helped me a lot and helped me get into the new businesses."[143]

GROWING PAINS

When last we left Don McDaniel, he had four plants—the parent plant in Sebring, Ohio, one in Peoria, one in Nashville, one in Charlotte. When MPI's largest customer moved to Texas, McDaniel decided a Texas plant made sense.

You know, Texas is like a separate world. You don't buy stuff here in the United States, you buy stuff in Texas....The easiest thing to do was buy a plant. So we went down to the Dallas/Ft. Worth area and found about twenty little label companies....Five allowed us to talk to them....They all had visions of grandeur and wanted seventeen times their earnings. [MPI decided to rent space.] So we sent four people down from Ohio—a salesman, pressman, manager and a slitter operator. And today only the pressman is still there. All the others came back. Anyway, we were able to pick up an ex-Lord Label guy who is now our general manager [Phil Henning, III]. He does a great job in Texas, actually Texas has become our second best plant because of everybody's efforts.[144]

By 1984 it had become obvious that the Peoria plant was not well-located; then another small label company in Monee, a southern Chicago suburb, came up for sale. It had been owned by Ray Fagen, who had earlier worked for both Adhere and Label-Aire in Cincinnati. But Ray had died and his daughter "didn't really want to run a label company."[145]. She had told the Mark Andy Company she wanted to sell and they called around to let people know. Don McDaniel went up to see the operation. Although the plant wasn't usable, MPI bought the operation and combined it with the Peoria company. Bud Gray, ex-Fasson, ex-York Label and ex-Apollo salesman was asked, "How would you like to run our new Chicago plant?" He said, "Boy, I'd love that." "We got into Chicago because of Fagen Label."[146]

WHEN YOU'RE READY

One of the MPI employees was from California and had told Don McDaniel that "when you're ready for a plant in California, let me know, 'cause I'm from California. You know I'd like to go back to California."[147]

> *Just about a year later we decided to give California a try. So we sent him out and I said, "I don't want to be in southern California—too wild a place, too wild and weird. Let's try in someplace like Sebring. [I] thought someplace in northern California, not too far from San Francisco, but...out of the high rent district.*[148]

The chosen location was Stockton, California, and it is now MPI's third most successful plant.

BABY IN CONNECTICUT

In 1987 McDaniel received a phone call from an old friend, Dick Doran. McDaniel kept running into Doran, one of the people who started the Label-Aire distributorship operation. Doran owned Doring Label in Danielson, Connecticut. McDaniel was always saying, "If you're ready to retire, let me know, especially now we're out in California and your company is close to Boston, [that] would make MPI coast to coast."[149]

Up until the time of the phone call, Doran said he would never retire. Then, knowing he was ill, Doran did offer to sell. Doran's death within a year left the operation manager-less.

It was a small plant and it had four presses and fifteen to eighteen employees. They had this one little gal who was in her mid-twenties, I guess. She was the oldest [most senior] employee they had. She was a smart little gal [whose] name was Lisa Caron. Lisa kept saying, "If you give me the job, hey, I can do this." We saw how ambitious she was and all the things she could do...All the people knew her, liked her, worked with her and we made her the manager. She had...a baby last year. But really she ran the office by phone....The time she was off she would call in. In fact they had the biggest month [ever] the month she was off having the baby. They did that to show her that they could get along while she wasn't there.[150]

This gave MPI access to the Boston/New York market. An added benefit was that Doring Label was a Label-Aire distributor. "We have [the distributorship] in every plant but Tennessee."[151]

NEVER HAD SO MUCH FUN IN MY LIFE

The Connecticut operation was acquired in 1988 and McDaniel is on the acquisition trail once again. There are markets he would like to serve directly, such as Seattle, Denver, Kansas City, Minneapolis/St. Paul.

KLEIN: Would you do it all over again?
McDANIEL: Of course. Why not? I've never had as much fun in my life as I've had these last twenty-five years.[152]

EXCEPTION TO THE RULE

Craig Buchanan is one of the exceptions to the rule about moving around in the industry (getting basic training in one company, then moving to another within the industry). While attending college Buchanan took a summer job at MPI and, after several years in each of several departments, is now vice president of operations at MPI Corporation. He, too, is still very excited about the fabulous potential of this industry and the continued growth pattern for the future.[153]

EIGHT PRESSES, NO BUSINESS

After eleven years in sales and marketing for the famous check printer, John H. Harlan Company, Pat Patrick felt the urge to own his own business.

What he found was a bankrupt PSARL business with eight printing presses and a 30,000 square foot building, but almost no active customers. Since bankruptcy records are public, the customer list was available to Patrick, but—to his knowledge—he never called on any of them. Later on, however, some called him.

As the company neared bankruptcy it had cut as many corners as possible, losing most of the original customer base. Even those who remained didn't really want to do business with the company at all. First Patrick changed the name to Label America and decided to specialize in data processing labels.

> *All the soothsayers [said] it would be one of the fastest—if not the fastest—growth areas in the pressure sensitive label industry. Another reason was that it fit better in with my own background in the check business.... Data processing labels are really more related to the forms industry...than they are to the label industry in a lot of ways.*[154]

But the equipment he had purchased had some drawbacks. None was able to fan fold or perforate in-line. Soon Patrick was able to buy a sixteen inch Mark Andy 4120, a four color press with in-line perforating and fan folding. Then Label America was able to compete for larger orders.

Again Patrick listened to the soothsayers who predicted the demand for laser printed labels would grow faster than the demand for continuous form labels for dot matrix impact printers.

> *And we developed a line of labels that would process through laser printers and at that time [became] one of only four label companies [for laser printer labels] in the country.*[155]

Avery had been the first, Uarco, the second and Label America and National Printing Converters, third and fourth.

> *That was significant because that was the first time in our business career that we were truthfully able to stand up and say we are a leader, we're here first...and we really were able to capitalize on that.*[156]

Currently about seventy percent of Label America's business is in data processing labels with slightly more labels for laser printers than in continuous forms. Thermal printing labels is also a "high growth area," Patrick says.[157]

Patrick credits a competitor with helping him understand much of the business. Jim Smith (profiled earlier) of Tag and Label Corporation in Anderson, South Carolina, had occasionally farmed out some specialized labels to the bankrupt company Patrick acquired. This is the same Jim Smith who, when he took over the label printing business he and a friend had started, was scared to death because he didn't know how to run a business. History proved that Smith was a quick study. Patrick notes:

> *We just developed a real important friendship and he taught me a great deal [about] business in general and our industry in particular.[158]*

One other thing the defunct company had supplied (besides a warehouse full of red Day Glow grabber labels) was a tiny work force of five employees eager to keep on working. It was a good thing, admits Patrick, "I don't know where the button is to turn a press on or off."[159]

That's rare in an industry where most owners are proud of dirty fingernails, but the EDP label market is quite different than—say—prime labels. It's a niche, and Pat Patrick has been able to grow from no business to more than ten million dollars in a dozen years.

FROM CANNERY ROW TO MODULAR MACHINES, PART II

Bob Scheerer left Dennison in 1978 when he and his partner and associate, Paul Morgan, got together and designed a line of machinery strictly dedicated to pressure sensitive applications, Syncro-Motion. Morgan, says Scheerer, has a tremendous background doing one-of-a-kind custom packaging machinery.

> *So he and I worked together and developed a line....So we set that up and they [Dennison] were the supplier and we just did the detail work. I did all the concept work for them and we worked together and...Dennison had exclusive distribution rights.[160]*

What does Scheerer consider his major contribution to the industry?

[It was] primarily in the machinery element....We have a very successful business on a basic concept and premise....All of our machinery is modular....We don't have what I'd call a generic machine. All of our machines basically go out of here in effect customized to the label or the product. So it's been a totally different marketing...and sales approach. We just had Kodak in here, the Denver plant, and I'm doing six machines for them and all six put labels on, but all six are different...yet they're all the same basically.[161]

Scheerer has sold Syncro-Motion Corporation to a young entrepreneur looking for a company after selling the one he owned. Scheerer will consult for a time. Then—

The thing that's most exciting to me [in the future]—and I may have the wrong name, anyway—we have an executive Peace Corps in this country. I've been contacted by them and asked if I would consider joining. I've got a tremendous background in this to build plants. Life is kind of interesting. [I would be] virtually building a [food processing] plant every year.[162]

U2 TO UARCO

After graduating from Penn State in 1972, Paul Yeagle joined the service and went into electronic reconnaissance for naval intelligence. He operated for four years in England working with U-2s and B-66s. He ended up doing research at Penn State, so he took advantage of that and picked up an MBA.

He spent ten years with the graphics department at American Standard (plumbing and fixture company), which, somewhat surprisingly, does a wide volume of printing. In 1981 American Standard began looking at divisions which it felt were ripe for selling. "I decided," says Yeagle, "I'd better get out of this business."[163]

I had heard through a friend that this company up in York, Pennsylvania, York Tape and Label, was looking for a VP in manufacturing....I didn't realize it would happen so quickly. It was my first real introduction to flexography.... I had never seen a flexo press prior to that.[164]

York had been established in 1948 by Ren Smith, who was the man who hired Yeagle. Although Smith was a relatively young man, he died of a heart attack in 1983. His family was not interested in continuing the business, so, within six months, it was purchased by Uarco, the large business forms manufacturer in Chicago.

NOT SO ARTSY-CRAFTSY

Four years later the president, Jack Gingerich, retired and Yeagle took over. It was also about this time that the technology improved so dramatically. It became less artsy-craftsy, less dependent on:

A person running the equipment that was extremely talented and extremely knowledgeable and had years and years of experience. Nowadays I don't think you need those years and years of experience, you need somebody with talent and some pride, but the technology is there to give them the tools they need...to make a product that looks very good....That evolution has really accelerated in the past five years....It's a very rapidly evolving business...more and more scientifically oriented, discipline oriented and less and less an art.[165]

Yeagle says people are always looking for label companies to buy. "There is no other graphic art business that exhibits that kind of growth potential."[166]

MY NEIGHBOR, THE BUYER

In 1985 Don Buchta and his partner, Bob Dunsirn, decided to sell Mid-America. They each had two sons in the business. Rather than try to pass on the company, they both decided to sell it instead.

It was time to let the boys fend for themselves. They were all stockholders in the business so they came out pretty well. My next door neighbor was chairman of the board [for a corporation that made corrugated boxes] and when he heard that we were looking for a buyer, he said, "Well, Don, why don't we look at you?" So they got into the label business.[167]

Mid-America, now part of Menasha Corporation, also purchased a plant in New Jersey and one in Colorado to take a major position in the PSARL industry. All four of the partners' sons have continued to work in the pressure sensitive industry. Don's son Jeff is now general manager of Mid-America. Buchta says he has enjoyed a satisfying career:

> *I was pleased with our development, I was pleased with our positioning in the industry, I was pleased with the quality of the innovations we brought to the industry, very pleased with the cadre of people [who worked] for us. I don't know if I would have done anything differently.*[168]

CHRONOLOGY OF APPLICATIONS, 1980-1990

1980	Premium/Coupon Labels
	Squeezeable Labels
	Transdermal Medicant Labels
	CD Disk Label
	Direct Thermal Printed Labels
	Laser Printed Labels
1985	Postal Service CFSII/Mums Labels
1986	Volume Primary Labels
	No Label Look
	Thermal Transfer Printed Labels

CHRONOLOGY OF INVENTIONS/ INNOVATIONS, 1980-1990

1980	Laser Engraved Ceramic Anilox Rolls
1981	Integrated Computerized Labeling/Marking Systems
1983	UV (Ultra Violet) Inks
	Hydraulic Die Pressure Controls
1984	Electronic Controlled Flexo Presses
1985	Computerized Ink Matching
1987	Rotary Screen Press
	Combination Flexo/Letterpress/Screen Presses

THE MAN WHO WOULDN'T QUIT, PART SIX

Harold Scherer liked working for Westcote. He got expenses and a car a year but felt he was somewhat underpaid. In 1987 (the $2,500,000 sales year) he left. "I was a little stubborn and they were a little stubborn and so we just parted company." [169]

After he left Westcote and moved to San Diego a company called PIC Industries made Scherer an offer which he accepted. His job was to drive up to the Los Angeles area one day a week and call on his accounts. "What they really wanted was my accounts."[170] And they got them.

[1]English interview, p. 15,16.

[2]*Ibid.*, p. 22.

[3]*Ibid.*, p. 24.

[4]Bruce Bell interview with Bill Klein, p.5.

[5]*Ibid.*

[6]*Ibid.*, p. 6.

[7]*Ibid.*, p. 7.

[8]Joe Weber interview with Bill Klein, p. 5, 6.

[9]*Ibid.*, p. 7.

[10]*Ibid.*, p. 13.

[11]*Ibid.*, p. 17.

[12]*Ibid.*, p. 18.

[13]*Ibid.*, p. 17.

[14]Bill Klein interview with Bruce and Virginia Ronald, December 7, 1993 (a coincidence).

[15]*Ibid.*

[16]*Ibid.*

[17]Mauritz interview, p. 10.

[18]*Ibid.*, p. 11.

[19]*Ibid.*, p. 12.

[20]*Ibid.*, p. 13.

[21]Lon Deckard interview with Bill Klein, p. 2.

[22]*Ibid.*, p. 4, 5.

[23]*Ibid.*, p. 5.

[24]*Ibid.*, p. 5, 6.

[25]*Ibid.*, p. 6.

[26]*Ibid.*, p. 8.

[27]*Ibid.*, p. 16.

[28]Bob Dillingham interview with Bill Klein, p. 1, 2.

[29]*Ibid.*, p. 6.

[30]*Ibid.*, p. 8.

[31]*Ibid.*, p. 9.

[32]*Ibid.*, p. 10.

[33]*Ibid.*, p. 12.

[34]*Ibid.*

[35]*Ibid.*, p. 13. The HP3D is a duplex laser printer that will print both sides.

[36]*Ibid.*, p. 15.

[37]Paul Dunphy interview with Bill Klein, p. 2.

[38]*Ibid.*, p. 3.

[39]*Ibid.*, p. 7.

[40]*Ibid.*, p. 9.

[41]*Ibid.*, p. 17.

[42]Rüesch interview, p. 13.

[43]*Ibid.*, p. 14.

[44]*Ibid.*, p. 16.

[45]*Ibid.*, p. 17.

[46]*Ibid.*, p. 20, 21, 22.

[47]Allan Prittie interview with Bill Klein, p. 2.

[48]*Ibid.*

[49]*Ibid.*, p. 4.

[50]*Ibid.*, p. 5.

[51]*Ibid.*, p. 6.

[52]*Ibid.*, p. 6, 7.

[53]*Ibid.*, p. 8.

[54]*Ibid.*, p. 9.

[55]*Ibid.*, p. 24.

[56]Styers interview, p. 18.

[57]*Ibid.*, p. 19.

[58]*Ibid.*, p. 22.

[59]John Egbert interview with Bill Klein, p. 2.

[60]*Ibid.*, p. 3.

[61]*Ibid.*

[62]*Ibid.*, p. 6.

[63]*Ibid.*, p. 7.

[64]*Ibid.*, p. 8, 9.

[65]*Ibid.*, p. 10.

[66]*Ibid.*, p. 15, 16.

[67]Jim Sauer interview with Bill Klein, p. 1.

[68]*Ibid.*, p. 7.

[69]*Ibid.*

[70]*Ibid.*, p. 6.

[71]*Ibid.*, p. 9.

[72]*Ibid.*

[73]*Ibid.*, p. 10.

[74]*Ibid.*, p. 19.

[75]*Ibid.*

[76]Leo Benatar interview with Bill Klein, p. 1.

[77]*Ibid.*, p. 1, 2.

[78]*Ibid.*, p. 3.

[79]*Ibid.*, p. 4.

[80]*Ibid.*

[81]*Ibid.*, p. 10, 11, 12.

[82]Dick Schwartz interview with Bill Klein, p. 3.

[83]*Ibid.*, p. 6.

[84]*Ibid.*, p. 7.

[85]*Ibid.*, p. 10.

[86]*Ibid.*, p. 11.

[87]*Ibid.*, p. 11.

[88]*Ibid.*, p. 15. Lord Label and several other companies have allowed key employees to participate in the company's growth program.

[89]*Ibid.*, p. 16.

[90]*Ibid.*

[91]*Ibid.*

[92]*Ibid.*, p. 22.

[93]*Ibid.*, p. 25.

[94]Patrick Masterson interview with Bill Klein, p. 3, 5, 6.

[95]*Ibid.*, p. 5.

[96]*Ibid.*, p. 18, 19, 22.

[97]*Ibid.*, p. 15.

[98]Bill Klein quoted in Masterson interview, p. 15.

[99]Gray interview, p. 16.

[100]*Ibid.*

[101]Gray interview, p. 13.

[102]Norman manuscript, p. 14.

[103]Trungale interview, p. 5.

[104]Beaudoin interview, p. 17.

[105]*Ibid.*

[106]John Patel interview with Bill Klein, p. 1.

[107]Dirk Desanzo interview with Bill Klein, p. 5.

[108]*Ibid.*, p. 4.

[109]Patel interview, p. 4, 5.

[110]Motter interview, p. 10.

[111]*Ibid.*, p. 19, 20.

[112]*Ibid.*, p. 20.

[113]Desanzo interview, p. 8.

[114]George Corbett interview with Bill Klein, p. 11.

[115]*Ibid.*

[116]*Ibid.*, p. 12.

[117]*Ibid.*, p. 14.

[118]Desanzo interview, p. 11.

[119]Patel interview, p. 7.

[120]Gary McKinney interview with Bill Klein, p. 2.

[121]*Ibid.*, p.6.

[122]Company literature and telephone interview, R. J. Klein to Bruce Ronald.
[123]Piggyback refers to pressure sensitive constructions that have two release coated liners, two layers of adhesive and a face material which allows a label to be applied, complete with backing for future or further application (from the TLMI Glossary). The most common example is the IRS mailing label which you remove to place on your income tax return. Other specialized piggyback constructions are possible.
[124]Motter interview, p. 11.
[125]Desanzo interview, p. 15.
[126]Motter interview, p. 13.
[127]Williams interview, p. 9.
[128]*Ibid.*, p. 10.
[129]*Ibid.*
[130]*Ibid.*, p. 13.
[131]*Ibid.*, p. 15.
[132]Gray interview, p. 14.
[133]*Ibid.*, p. 18.
[134]*Ibid.*, p. 22.
[135]*Ibid.*, p. 26.
[136]*Ibid.*, p. 28.
[137]Alan Percival interview with Bill Klein, p. 4, 5.
[138]*Ibid.*, p. 8.
[139]*Ibid.*, p. 11.
[140]Lance Hailey communication with Bill Klein, p. 3.
[141]*Ibid.*, p. 4.
[142]*Ibid.*
[143]*Ibid.*, p. 5. Lance Hailey died after a long illness. Many in the industry remember him with respect and affection.
[144]McDaniel interview, p. 29, 30.
[145]*Ibid.*, p. 30.
[146]*Ibid.*, p. 31.
[147]*Ibid.*
[148]*Ibid.*
[149]*Ibid.*, p. 33.
[150]*Ibid.*, p. 34.
[151]*Ibid.*, p. 35.
[152]*Ibid.*
[153]Communication from Don McDaniel, March 1994.
[154]Pat Patrick interview with Bill Klein, p. 6.

[155]*Ibid.*, p. 8.

[156]*Ibid.*, p. 9.

[157]*Ibid.*

[158]*Ibid.*, p. 12.

[159]*Ibid.*

[160]Scheerer interview, p. 31.

[161]*Ibid., p.* 37, 38.

[162]*Ibid.*, p. 36.

[163]Paul Yeagle interview with Bill Klein, p. 4.

[164]*Ibid.*

[165]*Ibid.* p. 7, 8, 9.

[166]*Ibid.* p. *18.*

[167]Buchta interview, p. 17, 18.

[168]*Ibid.*, p. 18.

[169]Scherer interview, p. 6.

[170]*Ibid.*

SECTION SEVEN

IT KEEPS ON GOING...
AND GOING...
AND GOING...

1990s

As the industry enters the 1990s, very high-speed automatic labeling demands equivalent supplies. Non-nickable liners are extrusion-coated to avoid die nicks and processing downtime. "The first popular grade was HP Smith's 44PP, polypropylene extruded onto natural Kraft, and silicone coated."[1] Al Norman believes that "the emergence of silicone coating release technology coupled with high-speed label dispensing equipment, are probably the two leading developments in accelerating industry growth."[2]

New hot melt adhesive polymers recently developed to compete with Shell's Kraton have given hot melt adhesives additional features for PSARL applications. These include Exxon's Dexco polymers and Nippon's Zeon. Both emulsion and hot melt adhesives are continuing to grow in popularity, partly because of the tightening EPA pressures and increasing raw material costs of solvent based adhesives.

New formulated inks allow "excellent flesh tones in the prime labeling area," and "today you can't tell a water-based ink from a solvent ink."[3] MPI's Bud Gray is glad to see the PSA label manufacturers as "full-fledged members now of the packaging fraternity."[4]

Gray also notes that the fastest growing market segment in the early 1990's is the development of film products. Film products for squeezable type application and for what is called the "no label" look. Some of the brightest films have been polyesters and polypropylenes. DuPont has been a leader in film development, as has 3M in film packaging.

Jim Smith of Tag & Label Corporation certainly knows about the cooperation:

> *We have one product that is particularly interesting and it was jointly developed by us, DuPont and 3M. It is a product that is used on outdoor transformer—electrical transformer industry—a seventeen year life outdoors....It's a vinyl with a special adhesive developed by DuPont, and then what's called Tedlar on the top of that to help, and we worked with our ink suppliers to develop long lasting high graphite type ink.*[5]

Tag & Label Corporation is a highly technical converter with much emphasis on computers. Smith points out that such technical service is important:

> *Two and a half years ago we were [the smallest of] more than 37 vendors supplying DuPont Fiber Division...ten major facilities with about 110,000 folks working at those facilities....They made a decision that they would reduce from 37 to one....We're doing that work now....[We were the] only one of these 37 [that]could do everything they needed.*[6]

Smith believes this shows both diversity and technical abilities. Tag & Label Corporation has had phenomenal growth the last three years, 22 percent, 31 percent and "right now we're projecting about 30 percent growth, but we're only half way through the year."[7]

Another new development, this one in dies, is chronicled by Lee Carlson of Preston Engravers:

> *The flexible dies which we call TPD, or thin plate dies, have really revolutionized things. When they first came out fifteen years ago they were only capable of cutting data [processing] type materials. Now they can cut film, foils, over laminates—pretty much everything. We have three grades, black, silver and a gold. The gold grade TPD will give you nearly as much life as a medium hard engraved die at about one-third the cost.*[8]

OPPORTUNITY/CHALLENGE/GROWTH

The proliferation of PSARL companies in the last five years has meant that the variety of items produced has grown as well. Bud Gray points out:

> *Right now the business still consists of marking, identification, warning labels...ancillary or add on labels, and primary labels. You also have functionals that serve as seals on containers. You've got coupons and products like that.*[9]

The industry has responded to the ever-changing needs of the end users, and, as various converters concentrate on specialties, they leave space for new small companies to get started. Although growth is usually not 25 percent a year any more, "a more viable growth rate would be a nine to fourteen percent growth area."[10]

Mark Wert (Soabar, MPI) also points out that not only are the converters more sophisticated, their customers are also:

> *Thirty years ago you had to teach people what is pressure sensitive. No longer do you get that question...I mean I carry cheat sheets with me, I carry a sheet...that explains what flexo printing is, what letterpress is, everything, and not only that, but what is a die, what is a press....And prep. Prep has definitely gotten better than it ever was before....If you're telling someone you're delivering something in eight weeks, you've already lost the business.*[11]

The retail folks are now saying they don't have the personnel or time to do it [put on labels themselves]....They are saying to Mr. Manufacturer, you want to do business with me, you send that baby in completely labeled, completely identified.[12]

A GOOD LIFE

Bill Eiseman, U.S. Tape and Label, says he doesn't have that big a plant...

> *We have 30,000 square feet, but we do a lot in here....It's a nice business. My son loves the business. I guess he sees that I like it—that's why he's here. We had fifteen percent growth this year. It keeps going....It's been a good life, a good business. You have many rewards.*[13]

ICE CREAM MAKES PEOPLE FEEL GOOD

There has been a Zimmer in the packaging business since 1930. Bread wrappers kept the company growing throughout the Depression and Zimmer made wrappings for bearings during World War II.

Karl Zimmer II joined his father's firm in 1964 after years in the publishing business. At that time the company was making pressure sensitive novelty stickers for bubble gum packages. This led to literally billions of in-package premiums for Kelloggs, produced roll-to-roll, but delivered to a lithographer for printing in large sheets.

Today Zimmer Paper Products Incorporated is the largest single supplier of popsicle and ice cream novelties in the country. "Even when the economy was really suffering," said Karl Zimmer, "ice cream makes people feel good."[14]

Pressure sensitive is about 35 percent of Zimmer Paper Products' business, with packaging supplying the remainder. Karl's son, Karl III, is now president. Another son, Erik, has recently joined the firm. Erik comments on the pricing pressures facing the industry today:

> *Let me give you an example of something that boggled my mind. Airline baggage tags. The old tag cost about a penny, a color-coated tag with a string around it....I think it was about ten square inches. The new baggage tags are 39 1/2 square inches...thermal direct...printed and die*

cut on both sides, and they only want to pay a penny and a half for that tag. Three times the material and you're also talking about pretty sophisticated material. That synthetic thermal is an extraordinarily expensive sheet. By the time you get through doing it, there's absolutely no money left in it. The liability is massive.[15]

Improvements in the bar code readers are lessening the need for such super high quality sheets for thermal transfer. This could change the EDP label business.

Zimmer is more closely aligned to European technology than most of the firms interviewed for this book, and firmly committed to hot melt. Said Karl, "We're matching silicones with the adhesives, so now you can get just about any property you want. The range of hot melt adhesives is very wide."[16]

The Indiana firm completed its third record year in 1993 and has done so by shying away from the commodity marketplace. Said Karl:

We don't do milk jug labels, or much EDP or anything like that. We've sort of aimed towards piggybacks, thermal transfer, direct thermal and the impact kind with NCR [carbonless] papers—really niche market.[17]

SECRETS OF SUCCESS

Andy Beck of API Graphics says the secret is "the selection of customers."

We're very careful about who we choose to do business with. That sounds kind of smug, but there is a lot of good business out there and [there are] a lot of good customers. We think the key is to let our competitors deal with all the bad customers and we'll take all the good ones for ourselves.[18]

COMBINATION PRINTING - GREATER GROWTH

"We see the significant thing today in the market we serve is combination printing."[19] That's what Package Products' Skip Heintzelman said. As noted earlier, Package Products is part of Engraph.

We print by flexo, by rotary letterpress, by flat screen, by rotary screen and by combination of rotary screen and rotary letterpress. Right now we do a lot of work that is combination screen, rotary letterpress and hot stamp in line. Obviously if we continue on we make the graphic opportunities broader and broader, which means that the label industry can grow more and more.[20]

Growth is something Heintzelman knows about. When he joined Package Products in 1962, the company was doing a very respectable four and a half million dollars. Today Engraph, the holding company for Package Products, is doing $210 to $225 million!

Before quoting Heintzelman and his views concerning the state of the industry, a quick look at Heintzelman's career. An engineering management major, he worked nearly a decade in the rubber business before joining Package Products. He was even in on the resealable bread label application, selling 31 applicators to a twelve bakery operation in Texas. In 1972 Gulf Oil was sponsoring the Republican and Democratic conventions and came up with a clever promotion involving state flag bumper medallions.

They were printed in eight or nine colors and were delivered as cut pieces. There was one for every state with the information about the state on the back....At the time the business was awarded, it was the biggest piece of pressure sensitive business that had ever been placed at one time—in excess of a million dollars.[21]

He was around when Engraph was created.

We got to the point where Package Products had made several small acquisitions and all of a sudden the top management people were spending all their time with the acquisitions, but...Package Products was picking up all their costs and expenses. We began to have trouble [knowing] who was the tail, who was the dog. So at that point we created the holding company called Engraph.[22]

Heintzelman believes that as graphic capabilities are increased, environmental concerns will prove very profitable for the PSARL business:

With all the environmental considerations, recycling, and solid waste, a lot of things that used to be [packaged] in folding cartons and all those good things, they're gone. The product stands on the shelf without a fancy wrapper on it and the label has to replace that.[23]

Europe is ahead of us on that and I guess Germany is the leader. [In Germany] you are expected to take the outer package back to the retailer and the retailer expects the distributor or manufacturer to take it from him....The label industry certainly has to be in a position not to end up being a culprit or a villain in this area.[24]

Heintzelman believes plastic face sheet (except for vinyl) can be used on plastic containers and recycles well. He has some very interesting views on the future of the industry:

The industry is re-stratifyingfying. I don't know what the exact numbers are...but there are somewhere between 1500 and 2500 people producing roll labels. Yet out of that group...ten to fifteen percent of that number generate 75 to 90 percent of the sales volume.... You're still going to have this extremely large cadre of small people who are local, do the small volume stuff, the straightforward stuff, the stuff you need two to three days' delivery on. But the stratification is becoming more obvious. I think in North Carolina we have six or seven roll label producers. The volume that we do is probably double the total of the rest of them. On the other hand, if somebody comes to us and wants us to do the same thing these other six or seven can do we're not in a position to do it competitively.[25]

The business, Heintzelman says, is still entrepreneurial in nature. He says the smaller companies who know how to run a business can still do well. (In fact, Engraph is made up of thirteen different business units, each a separate entity operating "as an entrepreneurial business."[26])

But the stratification continues, he says, because the equipment costs are so high.

I can remember fifteen years ago...when we bought a four to six color Webtron, we were paying $50,000 to $60,000 for that press. To buy the sort of flexo press that we buy today, we're looking at $650,000. Our growth in the past few years has been in screen and rotary letterpress and [of] the last four presses we bought, I think the cheapest was a million, two.... On the other hand, the press we spent a million, two for cannot produce 25,000 labels to go on the back of TV sets that says, "Call Joe's Radio and TV for service."[27]

STOP, THIEF

Al Sobel of Coronet Paper, who has been in the field since 1948, says the reason people fail in the pressure sensitive industry is:

> *The trouble is a lot of people stop. They stop pursuing the new idea or the new application. Because, believe me, they're out there....and crooks are increasing our business, or increasing the need for our business. We're developing a technique, through our customers, for buried bar codes. If you put a bar code label on the windshield of your car, there are people out there, believe it or not, who transfer that label or bar code or license sticker to another car. Or try to. And ours would immediately say* ***VOID*** *if you remove it.*[28]

One label that thieves steal is the state inspection sticker. The stolen sticker is then applied to a stolen or unregistered vehicle, or to one that could not pass the state test.

Thieves often use freon to freeze the labels and remove them. Even with freon, the VOID message appears. Another anti-theft device is burying the bar code in the adhesive. If removed, the bar code is destroyed. There are still thousands of applications waiting to be discovered.

WORKING FOR THE U.S. GOVERNMENT

Although John Egbert of McCourt Label turned down a government job, the postal transfer label, Aladdin's Dick Schwartz is even stronger in his opinion of the government.

> *[Someone] said one percent of the people generate the jobs. He's right. I'm one of the one percent. I got to keep outsmarting the government to figure out how to beat them. Because I'm not going to be a sheep in the flock.*[29]

But Darrell Dochstader of GarDoc says that security printing is in the forefront of PSARL developments:

Our biggest success is the anti-counterfeit device in the U.S. currency....The U.S. currency currently uses a security device that's printed by a company called Technical Graphics, which is a joint venture between a printing paper company and GarDoc. We also do passports, we did a lot of the stamp development for the government.[30]

THE MULTIPLIER EFFECT OF LEADERSHIP

Dochstader points out that GarDoc is, and has been, on the cutting edge of industry development. "We have a lot of very talented people here that work for GarDoc because they consider us a leader. So once you're up on the edge, you attract people who like the edge. It's a multiplier effect."[31]

Another reason for GarDoc's leadership, says Dochstader, is:

We have a 401K program that the insurance companies tell us is as good as Digital's.... We also pay probably five to ten percent above the current pay scale in the area...[and] since seven years ago...we work on the Japanese system....If you go through our plant now, [the systems are] probably more Japanese than American....Everybody is cross-trained now....We pay people for being really super generalists. And the net of it is that when somebody is missing, somebody from sales service goes out and runs the machines that day.[32]

But even GarDoc thinks it's only twenty percent of the way there.

Our objective for years to come....we're going to build a lot of our own machinery... since we've eliminated so much of our warehouse space. We don't need warehouses anymore...Then we'll have...a ninety percent system where a truck unloads, runs through the process, goes around the other side of the plant, picks up finished goods coming out the other side.[33]

Other GarDoc changes include empowering the employees to work on problem-solving teams. "These people are really going after this with a vengeance....We are giving them a budget and we're saying, 'Hey it's your job, so solve this. Here's how much money you have to work with, and if you need more, come back to us.'"[34] GarDoc anticipates no problems in meeting future governmental or even the ISO 9000 requirements.

COMPUTERIZED PRE-PRESS

One of the *avante garde* uses of the computer is in pre-press. Graphics are applied directly to the label from the computer, with no intermediate plate-making stage. Kalamazoo's Jim English says:

> *We start on the computer now with the graphics and we do the label's design and all these automatic step and repeats, distortion, chokes and spreads, out of the computer comes the film. And so the old artist's table, the green visors, and all that's gone....Now we take a model and change a dress on her, put long hair on her, lift her face.*[35]

Dale Bunnell of Mark Andy notes that...

> *We've got customers today that just have their computer linked to the converter's [computer] and the customer generates the artwork on a computer and transmits via modem directly to the converter without even a telephone conversation....The art work in-house...on a computer disk that generated the plates and it goes to press.*[36]

EPA COMPLIANCE

Today the labeling laws have proved to be a boon to the PSA label printer and, somehow, the cost to the consumer has proven to be minimal. This is just a start of further expansion of label opportunities due to federal and state regulations.

Compliance with EPA and labeling requirements today means many things. For Allan Prittie of Arpeco Engineering it's cation inks:

> *[There's] a major shift coming in the makeup of UV inks today...in my opinion a wonderful future because they have very low odor, very low taint, and so they are good...for food applications and they are in relative terms much safer....The cations are, in our opinion, part of the real wave now.*[37]

Prittie also is concerned with ink curing:

> *Solvent and water based inks...have environmental pollution problems, wash up problems...and the underground water table, they still can leach and obviously there are solvents going up in the air. UV curables are quite different in both areas. They don't leach, they're much more stable, there is much less environment pollution.*[38]

Some companies are currently more environmentally conscious than others, says Dan Tomlinson of Labelcraft Products, who cites, in particular, Jim English of Kalamazoo and Dick Schwartz of Aladdin Label.

"We've got two trends going on here," says Engraph consultant Lon Deckard. "The government mandate for less packaging, the government mandate for more information [on labels]. The immediate solution from my point of view is other variations of the multi-ply coupon."[39]

"There is a company ...in Washington State that...developed a way to chop up all these plastic bottles of mixed resins and somehow differentiate and sort out all the different resins and even with labels on them, and adhesives, too. So there is remarkable technology going on," says Bud Gray.[40]

TODAY'S LEADERS

Various points of view are expressed when questioned about the role of PSARL converters and suppliers in the 1990s. Gerry Gartner (GarDoc) suggests that...

> *the Engraph group people, (as did Spear, Inc. in Cincinnati), have certainly opened the screen business to primary labels. Well, Package Products and Patton, (Engraph), Avery certainly have been there. The end user focus that Fasson and some of the other stock suppliers have utilized has brought credibility, and the raw materials suppliers have done a good job at it. Mobil is doing a lot in that arena. So it's really a combined effort.*[41]

Gartner points out that many of the conglomerate companies are less niche-focused, and that there's not enough in any given niche to justify large corporations. He sees pressure sensitive as the preferred way of "decorating the container....Second and third generation changes in the pressure sensitive [labeled] container...every one of those changes is an opening of the door."[42]

Bud Gray at MPI notes that the trend for large forms houses for the last several years has been "to gradually begin divesting themselves of...forms due to slimmer and slimmer margins and buying up roll label houses wherever they can."[43] He believes that companies that decided...

they were in the packaging business, not the label business [are the ones which] continued to grow and grow....You've got two and 3 million [dollar] label converters who say they are in the label business, and you've got companies like MPI and others, who at 80 million dollars plus, lay claim to being in the packaging industry.[44]

The systems approach continues to grow.

MPI goes out to an end user and we come in with a systems approach. We not only sell automatic equipment, but we sell finished labels from one color to nine colors, offer them flexographically printed, letterpress printed, offset printed, and then we guarantee the whole business. And if it doesn't work, obviously the customer doesn't owe a dime.[45]

SMART PRESSES AND SMART PEOPLE

A smart press today is a computerized press. CAD (Computer Assisted Design), CAM (Computer Assisted Manufacturing) and computerized testing all make for more accurate, consistent products.

We all saw what the smart bombs did in the Iraq War. Well, the smart presses do the same thing. You set a job up with a computer, you have a CRT to look at, and every time you go back to that same job the computer would press up for you the same way, the same time, the same place, so that the greatest upheaval now is the new generation of printing presses.[46]

According to Ferdinand Rüesch of Gallus, the next step is the manufacture of rotary screen for UV flexo.

We can say that American crafts manufacturing industry has closed the gap to their European [counterparts] with UV flexo and UV rotary screen as of today. These new breeds of machines...have selected advantages and this is going to be the new evolution of the 90's. Take advantage of UV flexography, integrate the UV rotary screen and be able to satisfy all the needs now of the market for prime label-high quality four color process on foil paper.[47]

Rüesch also points out that the competition is no longer European.

Basically one of the leading manufacturers is obviously Comco, the newcomer in this business. They have the first real UV flexo UV rotary screen machine in a modular design. The company just introduced these technologies...and has shown it just recently at the Label Expo in Chicago and got...feedback as being the most advanced press in the world...Mark Andy's challenged, but now by an American manufacturer. European domination of UV letterpress and screen has disappeared now.[48]

One of the reasons that this is all attainable is that this business is now an industry. "It's global in its thinking. 'Let's go see what happens in Europe. Build some relationships with people in Japan.'"[49]

JAPANESE COMPETITION

However, Allied Gear's Jim Hart is impressed with Japan's Ko-Pack Corporation.

The Ko-Pack rotary letterpress [does] some amazing converting jobs. Now Ko-Pack is really a printer that sells some of its technology and I guess that's maybe the advantage they have over the American companies. Allied Gear and certainly, I think, Mark Andy, should be a little envious of Ko-Pack and some of the things they sell—it's technology. I think most of the American companies are like us, we're selling a piece of equipment, leaving it up to the converter to do what he will with it. Some of our foreign competitors will sell a converting system.[50]

THE CONSTANT IS CHANGED

Hart says the effort required to keep up with changing technology is becoming harder to do:

I think that change just comes ever faster. We see so many new packages, almost every new product now seems to have a new package. All of that has to have design behind it and machinery behind it....You just don't have time to investigate all of them some times.[51]

At the time of the interview, Allied Gear was doubling its plant size. Hart said he was looking at a souvenir hard hat awarded at the groundbreaking, "and let's see, we've got one, two, three, four, five labels on the hat. Three on the outside and two on the inside. Then on the inside there's a piece of foam [for] the support system, and that piece of foam was cut rotary, too."[52]

THE CANADIAN CONNECTION

When people in the industry mention the movers and the shakers in the PSARL business, they usually include CCL. The company, with plants in Canada and the United States, is a relatively recent comer in the field. It started in Canada when CCL acquired Meco and entered the PSARL field. In 1983 they purchased the packaging plants of Continental Can of Canada. Approximately at the same time they bought Modern Press in Sioux Falls, South Dakota, becoming an international company. Soon they acquired Anco in Sheldon, Connecticut.

Today CCL has two U.S. plants, in Connecticut and Sioux Falls, and a small corporate office in Grand Rapids, Michigan. There are two plants in Canada located in Burlington, Ontario, and Winnipeg, Manitoba. CCL's U.S. business is currently run independently of the Canadian operation, which reports to a Canadian president. The U.S. President, Bob Sirois, had worked at Continental Can for sixteen years and decided to stay with CCL after the takeover. He was made general manager of the eastern region—central Ontario to the Maritime Provinces.

In 1987 Sirois returned to Toronto as general manager of sales for Continental Can Canada. Within a year he was vice president of marketing and sales. Then he spent three years in an entirely unrelated field:

> *In 1989 I got offered a challenge that I simply could not put down...I was named vice president marketing and sales for the Canadian Postal Corporation....[It's] a Crown Corporation....You have one shareholder which is the government, but it's run as a business.*[53]

In 1985 the postal system was losing $600 million a year. For fiscal 1989, it made a profit.

We made a profit for three successive years....I stayed there for three years...and on January 1, 1992, came back with CCL. I felt I had done my thing with the post office. It was great, an experience one would never forget, but it's like doing service in the military.[54]

In February, 1992, Sirois was named president of CCL Label in the United States. Barry McKillip was in charge of the Sioux City operation and is now their vice president of production. McKillip believes converters, such as CCL, are driving the industry: "I don't see as much innovation on their [equipment suppliers] part as probably there should be....I would like to see [them] become a little more pro-active."[55] McKillip is optimistic about the industry's future:

You're talking about things like the no label look. And you're looking at other applications—holographic and security labels and all that which are basically film labels....So there are opportunities in packaging, EPA requirements for additional information on labels, all of these things cause turmoil. But in that chaos you can thrive.[56]

Is there a trend toward the multi-plant label producer?

I see certainly a continuation of some conglomeration. Probably I would have expected that it would have leveled off by now, but I would imagine there will eventually be ten major players....I mean there certainly is a lot of pressure because of packaging and the environmental concerns regarding packaging. I look for more consolidation and probably more of them [packaging suppliers] purchasing label companies to compliment what they have. But I still see the mom and pop shop as being very much a part of this industry.[57]

Sirois says CCL is currently (late 1993) repositioning its company, "saying that CCL is a label company, but they are [also] the packaging and communications company."[58]

Does that mean a partnership in packaging?

I hesitate to use the word partnership. Because I'll tell you I use that word very seldom and very carefully....People don't understand what it means....I'm seeing it starting to happen now. There is more and more pressure on our customers; we feel they really need more help and therefore there is an opening of the mind that for me is welcome, because you can be part of the process in a much earlier stage....When I came into this business they would say the label was the last component bought and therefore whoever can do it in two weeks [gets the job]. I see a major change in that. And I see much more planning. I see much more focus on the label side from the big companies....What I hear from large marketers now is that what they are looking for is energy. And that energy [is how we keep] coming up with alternatives, solutions, new ideas.[59]

Sirois makes one final point. "Service. Service. Service. And one way to get better at it is to improve the quality of your people in those areas."[60]

INTERNATIONAL NEWS/S.O.P. PSARL INDUSTRY

QDCS—Quality, Delivery, Cost, Service. The Japanese formalize their factory organizations around QDCS. Mark Andrews, Jr., agrees with the concept:

The quality college...the Malcolm Baldridge Award... all of these things head a company in the same direction and that is to delight your customer....So I think we did that in our industry in a very natural way in the early [days], not even realizing the total quality management.[61]

Flexographic printing may have begun in a small way in the rest of the world, but things are changing:

This recent Label Expo show here in the U.S. was visited by a convention from Japan of 25 owners of printing companies....They toured our factory...and their eyes are wide open to flexo. They just can't believe that flexo can print the quality that it's printing.[62]

Bunnell said Mark Andy's international sales now average about 35 percent of the firm's total business. Not only is flexo big in Europe,

It's virtually taking over Australia. It has been a good market for us. South America, Central America and Mexico have been tremendous markets recently. Even Japan is making a conversion to flexography. China, Thailand, Hong Kong...all those countries have already purchased flexographic equipment....We're fortunate to be in an industry that is going like this.[63]

Allan Prittie agrees with him. "I think we're going to give the Japanese and especially the Germans a very big run in the future."[64]

ISO 9000

ISO 9000 is the international standard for world wide quality control. Meeting the ISO 9000 standards implies meeting a universal acceptability of a level of quality agreed to throughout the world. Industry is setting world-wide standards which means that companies which wish to sell internationally must produce documentation, verification of specifications.

Today meeting those recognized standards is optional for companies in the United States, but as corporations with international offices and companies with international partnerships are learning, the ISO 9000 standards will need to be met in order to continue to do business.

The Malcolm Baldridge Award is often mentioned in the same breath as ISO 9000. Begun as a Japanese award for excellence in quality control, the Malcolm Baldridge Award is now world wide, a sign of excellence to which many American companies are beginning to aspire. Says GarDoc's Dochstader,

I think within a couple of years, we don't want to win the Malcolm Baldridge Award, but we want to be one of the runner-ups. Within a couple of years we'll be in a position to do that.[65]

A DIFFERENT POINT OF VIEW

John Garber's Label Technology is on the cutting edge of PSARL technology. After working with Mark Andrews in 1958 to develop the 5100 Mark Andy press, the UV cure litho press, Garber developed the machines into a seven inch, then a ten inch version. "One of them is still in operation at Keller Crescent right here in St. Louis."[66]

Garber believes in innovation. "Rotary letterpress is dead for all...purposes. They just haven't lain down yet....As soon as the plate technology comes up with a little higher durometer plate, there won't be any rotary letterpress."[67]

Garber built a 22 inch flexo press specifically for box board. "And the quality they are doing there, you couldn't tell it from offset most of the time."[68] He designed and built presses for Rand McNally that incorporate the consecutive numbering for airline tickets, and in California, bar code numbering presses that print parking tickets.

Garber likes what he does. "As soon as you get to a point where there are three people doing the same job, it becomes a *me-too* job. I'd certainly go through it again."[69]

FULL CIRCLE? MAYBE

And Garber is now experimenting again:

> *[There is a company called] Package Service Company in Kansas City [owned by] Wes Nedblake. He is truly an innovator in the industry....It has been his goal...to eliminate the liner and change the method to which the major producers apply labels to containers...a new labeling system for the converting industry. It will be a revolutionary change from what's been done over a period of years, because we're attempting, as everybody has for some years now, to eliminate the liner. I believe we're just on the verge of doing that. Some of it is in the patenting process right now....*[70]

The linerless label reverts back to the original concept of printing on self-wound tape. Nedblake is bringing that concept up to date.

THE LINERLESS LABEL

Wes Nedblake of Package Service said he never joined TLMI because, "I might teach somebody too much."[71]

That may sound like boasting but Package Service grew from $400,000 to over $30,000,000 in sales from 1979 to 1989, and that does not count the impact of linerless labels, which Nedblake says is revolutionary, not evolutionary in nature.

> *I was the guy who said fifteen years ago there will be ten label converters. Everybody laughed at me [because] they were popping up like gas stations then....There are 99 million mamma's and papa's out there still, and there are about ten guys who have the foresight and the technology, the capital and the people and resources on all fronts to be able to really offer the future in label technology.*[72]

As impressive as his recent growth has been, Nedblake believes it is only the beginning and said his patented linerless label will create explosive new growth in the industry. According to Nedblake, the 1990s linerless product is not the result of creating a new combination of old ideas, but "new combinations of new techniques."[73]

> *We're talking about a product that has no liner, that has a dry adhesive, activating the adhesive with an electron beam or with a strobe light at the point of application, and die cutting the label with a laser at the point of application. [This eliminates] rotary cutting dies in the marketplace and waste stripping the matrix and all the solid wastes.*[74]

Wes Nedblake believes there will soon be four or five major players in the industry, *that will have rotary letterpress, web offset, gravure, silk screen and flexo. Five systems that [will be] required to be competitive on all fronts in the marketplace today—and to provide all the service required by multi-national companies.*[75]

KINDERGARTEN FAILURE

Nedblake started in the business over 35 years ago. "I flunked scribbling in kindergarten, and I finished the eighth grade and got the hell out of there because I couldn't make any money. And I've been making money ever since."[76]

His parents had a small roll label operation in the late 1950s. Hallmark Cards was a client. His mother was primarily in the tape business, while his father organized a sales operation of perhaps fifty people who sold water soluble gum tape to locker plants in the food business. This started in the late 1930s. In the late 1950s his father and mother divorced and Wes worked for his mother producing labels on a machine purchased from Ever Ready in New York.

In the 1960s they switched to a Mark Andy label press (he still has the 1961 Mark Andy press), but it was only in 1969 they bought a 650 Webtron and began to produce PSA roll labels. In the late 1970s and early 1980s the business really began to explode, partly due to an improvement Wes made to the Andy 4120 label press. "Mark Andy couldn't figure that a press pulled paper instead of pushing it."[77]

Nedblake says he has 175 employees and they contribute to sales of about $188,500 per employee, about twice what Wes believes is the national PSARL average of $95,000 each.

He is particularly critical of the paper industry which he says is an industry designed to sell trees. With an ongoing switch to non-paper labels (mostly plastics), the marketplace has changed. Wes Nedblake plans to be in on the cutting edge of the new technology.

But he hasn't completely forgotten about the past. When told that Stan Avery was still playing tennis in the 1990s—with two hip replacements he said, "I'll be damned. God bless him."[78]

GEORGE AND THE COOKIE MONSTER

Would you believe a rotary silk screen press in 1949? Said George Reinke, Screen Printing Systems, Inc., "It was an abomination, but it did a job in its own way."[79] During high school back in the late 1920s Reinke worked part time in the silk screening business, full time after 1930. It wasn't pressure sensitive then; that came in the late 1930s. "Some of them [the labels] worked, some of them didn't. If they stayed on the shelf for two months you didn't know what would happen."[80]

First Reinke tried to build two color, thirty inch wide rotary silk screen presses, with an emphasis on printing wall paper. By 1970 Screen Printing Systems was creating "all kinds of specialties."[81] The most unusual was a contract with a Japanese outfit that makes cookies and screen prints images on them. Screen Printing sold the Japanese company about 25 or 30 screen presses and, although no new presses have been ordered, they still make all the new screens for the company. The presses...

> *print on the raw dough and bake them. [The] cookie might be two by three inches and they are usually about a 32 inch repeat, and about a meter wide. They turn a lot of cookies out. The Japanese are more the cute little stuff. Now we have a few presses here in the United States. They run these wrestling things on them. It's the cookie part of an ice cream sandwich.*[82]

Reinke sees hot melt inks as a wave of the future. He has patented an ink of his own. But...

> *UV is the big thing now. I've done a lot of work on the hot melts. I think sooner or later we get the hot melt inks to replace the UV inks. UV is a fairly hazardous material, whereas in hot melts, I made a hot melt... that's edible. That's what it's getting to now.*[83]

THERE ARE NO BARRIERS

Alan Campbell of Fasson, on the subject of the state of the art in PSA base materials today:

> *We spend a lot of money to make a product that is very consistent....We don't slow down to make splices, so there's not a change in process conditions. There's less variability in the adhesives we make...everything we do is designed to make a very consistent product. We have better moisture control and moisture does have a big impact that is kind of*

invisible. Our customers don't recognize it unless they have a problem....We understand the slitting process better. We have better tension control. We have less edge weave. We have more uniform moisture content both of the face and of the liner and there are fewer nicks so there are fewer opportunities to start a tear in the matrix.... The technology we have built over the last fifteen years really is critical in the design of new products.[84]

Ten years ago people thought the industry was mature and you couldn't get any better. We didn't think we could go any further. Now we've said, "There are no barriers."[85]

Fasson now links its own inventory systems to its customers' systems, monitoring a customer's inventory and producing replenishment stock which can be shipped without any specific order. Customers can also order electronically.

There is a whole new line of films for the plastics industry. They are polyolefin films replacing "PVC, which is both environmentally unsound and difficult to print."[86]

The printing of labels for desk top laser printers is a current development. Campbell says:

We've developed special products for laser printers. The adhesive is only part of it. You have to have the right thickness, it's got to have the [proper] release, it can't be too tight, it can't be too loose. It's got to go through a convoluted web path. There have been product after product developed. I think one of these days we'll have a coating line dedicated to making [laser printer] products.[87]

Thermal printing is another growth area:

Direct thermal printing is—I believe—still growing. Thermal transfer printing is definitely growing because everybody is putting bar code systems for controlling inventory into their factories. Not only for inventory but for in-plant control of the product as it moves through the product line.[88]

What about the global picture in the pressure sensitive industry?

It's complex. As the world becomes more global...there's more flexo going into Europe and Asia and more letterpress coming from Europe and Asia into the United States. Every printing technique has its own strengths and weaknesses.

There still are an awful lot of seal presses, mostly from Japan. They are throughout Asia in Taiwan, Korea, middle Asia, in Indonesia and it's in Australia. Australia has a lot of stuff from Europe. A lot of Gallus equipment. Gallus has been very strong in letterpress and UV letterpress and also in rotary screen.

Letterpress [rotary] has become popular in the United States because of its unique characteristics. That's been through Gallus and Ko-Pack. I think those are the two better-known names.

In Europe, speed has become important. In the early days they wanted very high quality printing and didn't worry too much about the speeds.[89]

Competition has changed. Now all over the world everyone wants both speed and quality.

DADDY, DON'T GO

Jules Farkas of Comco brings a unique perspective to the industry. A Swiss chemical engineer, Farkas started his working career at Gallus in St. Gallen in 1968. While working at Gallus as production manager he also worked on his Ph.D, which dealt with how manufacturing could be automated with computer technology.

In 1981 Farkas came to the United States with Gallus, Inc. At first it seemed like a bad idea. The Gallus rotary letterpress was three to three-and-a-half times as expensive as a comparable flexo machine. The quality was excellent, but great strides were being made in flexo quality at that time. Farkas recalled meeting Gerry Gartner of GarDoc at a TLMI meeting in Washington, D.C. Gartner told Farkas that the UV letterpress had only limited or no chance to make it in the United States.

This remark was not very encouraging for a young European salesman who was willing to conquer America...furthermore in 1981 the U.S. wasn't in terribly good shape with interest rates of eighteen to twenty-four percent....Nobody was buying.[90]

During a dinner conversation with his family, Farkas said he was thinking of going back to Europe.

Then my eight-year old daughter gave me the wisdom. "Daddy, we are here now. I don't want to go back. I like it here. So you have to come up with an idea. There's no other way, Dad." What an encouragement.[91]

Two days later a pharmaceutical company became his first sale.

THE FUTURE BY FARKAS

Farkas predicts great strides in press development in this decade. Interviewed in Spring, 1994, he said:

> *I think in the next two or three years we will see computerized presses entering the market to help reduce manufacturing costs to better meet just-in-time demands....The new presses will enable the printer to pick and choose between printing processes....There will be no compromise in quality—the quality of printing is a "given," defined by the customer. UV flexo, UV rotary screen, UV offset will be the dominant printing processes of the future, embraced by North American, European and Japanese converters. The printing presses of the future will be closely linked to the electronic pre-press and the presses themselves will, for the most part, be computer-driven. When we arrive at this point—I predict approximately 1998 to 2002—how far is it then to electronic printing?*[92]

Farkas says people are talking about PSARL as a mature industry. "People think it's matured because they don't have the right salespeople to sell."[93]

FASSON AIRFORCE

Chuck Reed loved inventing things, loved making Fasson base materials (which in the early days were of questionable consistency), loved to work on a customer's press, and really loves airplanes! When a customer had difficulties, Reed literally came flying to the rescue.

> *That was a losing proposition for me, too. I subsidized an airplane to [the] company for over twenty years. Now I understand they're using jets. Paying the whole bill.*[94]

Bill Klein on Chuck Reed:

There was a guy who could make something run, no matter what it was. He was a shirt sleeve type....I never remembered Chuck with his fingernails clean. He was the kind of guy who got in there and helped the guy make it work. He was a natural tinkerer. He had good chemical background in adhesives....He began to know presses, plates and everything else. Then he was almost killed in a plane accident in August of 1979. But he's all right now, he retired, he still tinkers with planes. In fact, he has a small airforce. He has a bunch of vintage airplanes on his farm.[95]

Some of Chuck Reed's aircraft, including a vintage World War I Junkers Fighter-Bomber, have been used in the movies.

ONLY IN AMERICA

Dick Capuzzo of General Trademark tells how Ken Felis broke into the business. Leon Beaudoin called Capuzzo and said he knew Felis was trying to get some used equipment. Felis bought a rewind table from General Trademark which was shipped to Label Systems, Felis' company, operating out of a defunct Pizza Clown restaurant.

Felis had worked for a label printer while finishing his CPA and decided he liked labels more than accounting. (He is one of several accountants to make this decision.) He had a small Webtron and soon started producing the mylar rings for computer floppy disks.

He went through some interesting changes there. They started with just die cutting these things and of course the prices dropped, and he decided that the only way he was going to do this was if he could set up his own adhesive [coating]. So he set up a coater. Then he set up a silicone coater. He got special dies made up and he had these special instruments for checking the consistency of these things and all.

A funny thing about the name of the company. The name came from Avery Label Systems. He dropped the Avery and used the Label Systems part. Anyway, Ken went on from there and got into a lot of interesting stuff. Today he's doing holograms....They actually have a hologram lab up there in his plant...the stuff you see on Visa cards and all that.[96]

Capuzzo's son is in General Trademark's art department and represents the fourth generation in the business. They have 27 employees, and one retired "a few years ago and he started with us in 1947....Another guy who's retiring this December [1992] and he started in '52."[97]

Capuzzo says he isn't in the PSARL industry for the money. With tongue firmly in cheek, Capuzzo says, "This is more of an ego trip for me than anything else."[98]

IT'S ONLY BIGGER, NOT DIFFERENT

This industry is bigger, not different. It requires that the entrepreneur [pay] attention to the customer, and one day you're the little guy knocking on the big guy's back door getting his business, and then all of a sudden there's a little guy knocking on your back door getting your business. (Bill Klein)[99]

Jim Hattemer, who sold Graphic Resources to Lon Deckard about 1974, but who remained with the company until its acquisition by Engraph, supports Klein:

You can still get started in a garage. To start out, [maybe] $150,000, less than a fast food business without as much hassle. [You] still can start small and grow it.[100]

Others are also in agreement:

Absolutely. Its a wonderful business if you want to work hard, and like you say, do the American dream.... You can buy a used Webtron for $50,000, put it in your garage and you're in business. If you want to make potato chip labels and that kind of stuff, you can make a decent living out of it. (Paul Yeagle, York Tape and Label)[101]

When we were starting out, we were a lot smaller. And if we were going to have to compete with the big guys like Avery, Dennison, we have to be [fast]. They'd give the service and their prices were usually high, too. Now we're there, we're there ourselves now. (Jim Wilson, MPI)[102]

Some disagree:

It used to be...the salesman from 3M would find a good pressman some-place and, bingo, they were in business. One guy got the orders and the other guy ran them at night. Sometimes it was the same person. I don't think you can do that any more....[You wouldn't have] capital, have the *bucks for a press now. We bought I think, probably the most expensive label press ever produced in the industry from Mark Andy and paid $45,000 for it. Now that wouldn't buy you an unwind stand. (Gerry Gartner, GarDoc)*[103]

But there are still some people coming in on a shoe string...

Well, the other thing is a guy can still get into the label business with $100,000, which is easy to borrow if you really want to get a label business. All you have to have is one salesman, one press operator, and you can turn out easily a million dollars a year with one or two presses. I see a lot of P&L statements, about forty percent, maybe a little more than that in raw materials, about twenty percent operating costs, and about forty percent gross profit margin out of which you take your salesman and yourself. (Jim Hart, Allied Gear)[104]

They're not our competition. They're the ones that are going to be selling mailing labels and pictures and stuff like that locally. (Gerry Gartner, GarDoc)[105]

WHAT'S NEXT? A LOT!

Lou Werneke sees a bright future. "These guys are all...positive thinkers. Just this summer I was talking to Belmark and Bruce Bell, who's a member of TLMI. He expects a thirty percent increase next year."[106]

One of the hottest things in our industry right now...is tinting inks. It's putting a coating on a sheet, it's got to be put on there right, because the sheet is porous. If you get a little bit too much it just makes the whole thing different. I found a guy that lives a mile and half from me up at the lake and he's worked for Chemolite's Coating Division for the past 35 years. I presented him with the [ink tinting absorption] problem....He said, "We go out and get a coater, then we take the whole thing apart and put it together the way we want to. And it has to be something you can duplicate and that Hallmark can duplicate and its branches all over the world....That's a big part of the growth of the flexo business.[107]

Werneke also predicts that presses like the ten inch Mark Andy, and narrow web printing presses with the pressure sensitive sheet will be producing..."inserts for General Mills and Pillsbury. Those presses [will] go 24 hours a day."[108]

Dirk Desanzo visualizes ten to twelve percent growth for the foreseeable future. "But you're going to see things like the pressure sensitive stamp come into play, which is going to expand the market."

Bud Gray points out that "this industry is not mature, by any means....It is vibrant, it is young. If I can relate it in terms of human growth, it's an eighteen or nineteen year old teenager about to jump into its early twenties."[109] The explosive growth of the industry seems to be a fact, as the equipment suppliers point out that forty to fifty new entrants a year is a reasonable number.

Innovation continues, such as the coupon type label with inserts. "It's probably only been in the last four or five years that you've seen these tremendous innovations,"[110] says Dale Bunnell.

Dan Tomlinson is cautiously optimistic. "There's too many people in the business, at least there is in Canada....I think the opportunities are going to be to people who look for niche marketing and find what they can do really well."[111]

Bill Klein agrees:

> *That was Stan Avery's dream...it was just like Martin Luther King's 'I have a dream.' And one of his [Avery's] dreams was a pressure sensitive postage stamp. [That] and a specialized roll that would handle baggy materials through the oven....That snake's skin roll that he worked on forever and ever. Of course, we now do have pressure sensitive stamps. In Stan's lifetime, and he's still chugging.*[112]

THE BEAT GOES ON

Percival Wise of Wise Tag and Label, who entered the tag business in 1928, echoes what a number of interviewees have said:

> *I will tell you that the PSA roll label business went far beyond what I thought it would in 1956. The quality of the product, the graphics that we can put on the pressure sensitive label today is so far advanced over what it was even ten years ago.*[113]

THE ULTIMATE LABEL

Avery Label, Fasson, the U.S. Bureau of Engraving and the U.S. Postal Service did come up with the first U.S. pressure sensitive postage stamp. Originally meant for Christmas cards, to save cancelling (the adhesive was permanent and the stamp could not be removed), the pressure sensitive stamp ran into problems. With no cancellation date, there was no way to prove when a letter had been mailed, for example, tax forms.[114] Pressure sensitive stamps are now common in Europe, and are becoming more so in the United States.

> *The ultimate label as far as we're concerned is the pressure sensitive postage stamp. That's gotten pressure sensitive down to as close as a mass market consumer product as I can think of. We made last year, I don't want to get my decimal points wrong, we made 1/2 trillion of them. So I mean there are a lot of them out there.*[115]

David Good knows what he is talking about. His company, Voxcom, is one of three primary manufacturers of PSA postage stamps. The other two are Bank Note of America and Stamp Ventures. Others produce them as well, including Avery and 3M, but the three are the current main producers. "Companies like Fasson, Kanzaki and Brown Bridge supply the material."[116]

YOU ARE YOUR OWN THREE MILLION DOLLAR CUSTOMER

"One of the top ten developments [in the PSA industry] will be development that is taking place here."[117] That is the opinion of Craig Daniell of Canada Coating and Laminations, a part of the Pridamor Group.

Daniell says it occurred to him there were about 3000 companies, many of them fairly small, who were buying their raw materials from four of five prime suppliers.

> *You have great big companies supplying tiny little converters. Everyone in the industry thinks of themselves as the supplier or the converter. There isn't a [TLMI] category for manufacturers. Now I consider myself a manufacturer.*[118]

What Daniell is doing is providing the converter with a forty inch wide machine that will coat both adhesive and silicone in line. Daniell estimates the converter can save about thirty percent by doing the coatings in house.

If you're running a label business doing five million bucks, you're buying three million dollars worth of paper, and if a salesman came in to see you and said, "Look, I have a new customer for you, a three million dollar order, you can make thirty percent on it." A label guy would get down and kiss his feet. Then [the converter] says, "Who's the customer?" The salesman would say, "You are."[119]

According to Craig Daniell, the coaters provided by Canada Coating and Laminations are inexpensive, fast, easy to maintain and require little in the way of floor space. He is especially proud of a recent start-up that produced only "two hundred feet of paper before we were in commercial product. Maybe $100."[120]

Daniell predicts major shakeups in the years ahead:

I didn't think you could write a book about an industry that is more ripe for change than the people who supply the pressure sensitive market. Every product has a product life and maturity. Every business has a maturity. Polyethelene film is still used, but you have to make your own if you want to be a player....Bill, I wouldn't want to be in this business today making labels if I could not laminate. I'll tell you why—because you aren't going to make any money.[121]

HOLD THE PRINTING

Although technically a label has to have something printed on it (or, as in the case of the first pressure sensitive labels, allow the user to write something on it), many PSARL firms have discovered other applications. Lee Carlson of Preston Engraving points out that Dr. Scholl's foot pads are done rotary, as are most adhesive bandages such as Johnson & Johnson's Band Aids. And the funny little patches for the nosepiece on eye glasses are produced by a rotary press.[122]

YOU WANT IT, WE'LL MAKE IT

From the motion sickness medication patch to today's nicotine patch, the transdermal patch business continues to make great strides. Dale Bunnell at Mark Andy was in it from the beginning:

> *I got a call from a company out in California a few years ago and the guy started the conversation with, "Well, you probably can't help me, but..." And he went on and explained this problem [motion sickness patch]. And I said, "I think that we can help you. I think that we have the equipment that will manufacture what you need.*[123]

Bunnell notes, that confidentiality aside, Mark Andy is working on equipment that will do medical testing...

> *your blood sugar, your level for stuff of that nature, bar code scanning that feeds into the computer the information so...they can't overdose or underdose the patient, because the computer monitors everything.*[124]
>
> *Credit cards. Lots of our customers do other than pressure sensitive, too...special promotions that go into cereal boxes, go on cigarette packs...and tires, all kinds of things.*[125]

And David Good of Voxcom says, "We can take that a step further. We're making a pressure sensitive in-line system [designed to coat, print rotary letterpress, flexo, offset, hot stamping and silk screen, die cut and strip]." He adds, "We've got machinery that we've custom designed and built ourselves."[126] And that's not all:

> *Look at the tapes for 8mm cameras...you'll see a hang tag that's on those. It's replacing the old blister card packaging....A graphic hang tag is the packaging cost that's been reduced and reduced....It eliminated the header bag and the plastic associated with that. And now these tags are made of recycled polyester. So they're not recycled, they are already on the second life....[The hang tag] was produced from raw materials in-line one hundred percent. And if you look at it you'll see that is the only way it could be produced, because of where the adhesive and where the printing is.*[127]

Like most of those interviewed, Good sees the technology "growing by leaps and bounds. We're a bunch of engineers and we're trying to position ourselves on the leading edge of technology and it's just tough keeping up."[128]

ON THE CURVE

Bill Whitfield is another who has positioned himself on the edge. As MPI's electronic prepress "guru"[129] he says, "I go around and try to disseminate the things I've learned and tricks I've picked up."[130] Since electronic prepress only started about 1990 and Whitfield began using the system in 1992, "I'm learning every day myself....You teach someone one thing and they'll teach you two things back if you listen."[131]

What it [electronic prepress] did is allow us to serve our customers better. We are customer driven, of course, and we're finding that numerous customers are starting to use...artists who have all figured out that they can buy a Macintosh and make it work. So we got to the point where we couldn't afford not to service those people, because we get a disk in here....We very rarely [get hard copies to work from].[132]

We wanted to stay competitive and be able to service customers....I don't think it was a dollar value....We have a lot of gourmet customized items—little jam and jelly companies that like five, six, seven color jobs, but want a run of 1,000 of each.[133]

Whitfield also has a challenge from a boss who is fact and figure oriented.

So what I did one night is...I walked into the dark room and I did every task I could think of, I timed them all. I took 100 jobs and I took them apart and built variables along with this, and darn if it doesn't work 95% of the time. It's incredible.... Electronic stripping on more complicated jobs doesn't necessarily help you go faster...it gives you more accuracy.[134]

Whitfield points out that this frees up time for creating things, because "the creative end is what takes time and energy, whereas the trapping end is more or less just the function."[135]

We have a policy...in this plant...and in other plants. The same artists that keep sending us stuff improperly done, we will educate them for free. Spend some time with them.[136]

All of a sudden the prepress came in and shot about twenty light years ahead of press improvements. So now what we're finding is the press is trying to catch up again....I'm a visionary because I'm trying to keep up with the technology in the computers for the prepress end.[137]

TIP OF THE ICEBERG

The view of the pressure sensitive industry by Chuck Miller, CEO of Avery, is much the same as that of the industry's founder, Stan Avery:

There are no negatives! This crazy sticky products industry is the most exciting business anyone could be in. I come into the office every day excited to hear about the latest new product development, and I'm convinced we're still only seeing the tip of the iceberg in terms of new applications.[138]

Some of the milestones that Miller remembers include the 26-inch coating machine at the Monrovia facility (1949), which allowed Avery to bring raw materials in one door and ship finished printed labels out the other. Permanent adhesive (1951) opened new doors for the PSA industry, and the Crack-n-Peel self-adhesive sheets were so successful "that the company had to create an allocation system to ration the supply."[139]

Like all the other successful PSARL executives, Miller understands that the needs of the customers drive the business.

Our commitment to giving our customers the best quality product for the best price in the fastest time as made us the flexible organization we are today—and we constantly try to do better.

It's all out there, a myriad of opportunities for new applications and profitable growth. This is a fantastic business....I'm grateful to have been a part of a rapidly growing, exciting business whose products touch all of our lives every day.[140]

THE WHOLE EVOLUTION

Beginning as a division of Colts Thomas in 1972, Spear, Inc., has spent the last ten years on the cutting edge of the pressure sensitive industry. Rick Spear, son of the founder, and his brother now run the business.

Although much of the business was with Proctor and Gamble, in the mid 1970s Spear moved into pharmaceutical printing, "printing plates, and always trying to do some very fine quality work that nobody else could do."[141]

It was actually a good foundation because although everything has come full circle now, everything's changed because now you are not supposed to do these things that are considered inefficient, you should have productions control the process....It's really interesting to watch the cycles...but back then we were highly acclaimed as having the best systems in the country.[142]

In 1982 Rick and his dad bought the business, incorporated it as Spear, Inc., and began focusing in on the screen area.

We got involved with [what] Fasson called Reveal™... doing Reveal with a binary code system, two different colors. We were the first company to bring that out after Fasson had developed it....Nobody had thought about putting polyester liner on a vinyl face stock. Basically that came about as Bausch & Lomb people developed it.[143]

Spear worked with Proctor and Gamble using pressure sensitive as a viable alternative for primary labels. P & G had a new polypropylene bottle which was squeezable and had a flip lid.

When it got wet and you squeezed this damn thing, it disintegrates [the label]....We came up with one stock...Flexcon's material..1/2 mil polyester laminated to 3 1/2 mil G liner with a super aggressive rubber-based adhesive.[144]

It took us about nine months [to get the labels working]...and we practically doubled [our sales] overnight. But we were losing money on it the first nine months.[145]

More Proctor and Gamble business came Spear's way as the corporation tried to come up with a more *cosmetic* approach to labeling. Spear pointed out that P & G was giving equal weight to the packaging, which had always been a second consideration. This forced Spear into trying new ideas, such as screen printing of clear films and hot stamping.

...the first ones done really in this country. There had been some done on a limited basis in Europe.... There were a lot of good people at Proctor...that were very pioneerish in their own way....Ed Fox...gave people working for him the freedom to get the job done....Today we're trying thinner and thinner gauges, 2 mil polypropylene, we're looking at 1 1/2 mil polypropylene, clear polypropylene. Back then, three or four years ago, they weren't even thought of.[146]

But Spear is not finished with innovation:

You could see the whole evolution, the process technology were we start off with flexo, then we go to screen and clear, then we add screen and clear hot stamping, and then you add combination printing, which became a big thing the last couple of years. We put letter press and now UV flexo...in line with screen printing and you back them up with white and you have pretty exotic graphics where you can hot stamp and do process work.[147]

The most recent development is ACL, applied ceramic labeling, where the ink actually becomes part of the glass. Starting with Clearly Canadian, the beverage industry opened up to this application, especially in the beer and ale field. But Spear is also the front-runner in APL, applied pressure sensitive labeling.

APL does work....We die cut right to the ink. You just practically don't know there is a label there. It looks ACL, it feels like it, but the graphics are much better. What is significant here is Anheuser Busch, the largest brewery in the world, has come out with the hottest product [ice beer] on a national basis, and they chose pressure sensitive as their method of decorating.[148]

How does Spear hold the edge? "We really listen a lot. We take a lot of risks, and we're long term thinkers."[149] It works for Spear.

THE PAST IS PROLOGUE

Isn't it amazing that a business begun with a borrowed $100 could grow to a $6.4 billion industry in less than sixty years?[150]

It's also amazing that most of the heavy-hitters today, including those recently retired, started with relatively low investments and are now rich men. If Horatio Alger were around today he would probably be writing about the poor little boy cleaning up the presses in a pressure sensitive roll label converting plant.

Stan Avery's tale is a classic story of entrepreneurship. Many of the people interviewed for this history follow the same scenario. Of those who did not, most are second or third generation operators/owners of a family-run business started by an entrepreneur, who grew up imbued with that same entrepreneurial spirit.

It's a business that has thwarted large and otherwise well-run companies which have tried to rule by committee. Unlike most companies that create a product and build it, the PSARL converter almost never settles into the rut of making the same product week-in and week-out. There are always new challenges which call for decisions to be made quickly. They must be the right decisions a high percentage of the time.

Obviously, not all the people in the industry are entrepreneurs, nor were all those interviewed for this book, but the most successful of them share the drive and the "Let's roll up our sleeves and solve the problem" attitude that characterize the entrepreneur.

From Stan Avery's modest start as the angel of the refrigerator loft to at least 6.4 billion in such a few years is incredible, but as Al Jolson used to say, "You ain't seen nothing yet." Most of the interviewees have experienced double digit growth for years and almost all anticipate more of the same for the foreseeable future.

Beginning companies often show dramatic growth patterns. Growing a $50,000 business by twenty percent a year is certainly not uncommon. But raising a 6.4 billion dollar industry by eleven, thirteen, fifteen, twenty percent a year boggles the mind.

And, although the industry is encountering some sophisticated challenges from both Europe and Asia, it remains an American success story. Stop by Hellriegels some time and hoist a toast to one of the most exciting industry success stories ever told.

CHRONOLOGY OF APPLICATIONS, 1990s

1990	Security Labels
	Instant Redeemable Coupons
	Multi Fold Labels (Expanded Content)
	Product Sample Containing Labels
	Two-Way Primary/Instruction Labels
1992	ATM Plastic Postage Stamps
	Hologram Labels

CHRONOLOGY OF INVENTIONS/ INNOVATIONS, 1990s

1990	Computerized Make Ready/Pre-Presses
1991	Recycling Release Liner into Fine Paper
1992	Matrix Recycling into Fuel Pellets
1993	Liner-less Pressure Sensitive Adhesive Roll Label Material

THE MAN WHO WOULDN'T QUIT, PART SEVEN

In December of 1992 Harold Scherer decided he had had enough of driving to Los Angeles, so he quit PIC Industries. Soon Scherer received a call from his old boss, Mort Allerdice of Westcote. He invited Scherer to lunch.

> *We [had] parted as friends and I kept in touch with him....I had lunch with him in January of '93 or late December, '92.... [Allerdice told me] "We have a man who is really too busy covering all of our accounts up in*

LA and in San Diego. How would you like to cover San Diego?" I knew most of the accounts down here. So they made me a proposition and that's what I'm doing today. They use me on Tuesdays.[151]

Scherer says he has about a dozen accounts and he would be going out tomorrow (the interview was on a Monday) to see if he could add another. But not too many:

I have enough to do. I play golf a little bit...so I don't want to work that hard. In fact I wouldn't want many more customers than I have now. [But] if a new company started up I would go after them....If half a dozen started up I wouldn't say that I wouldn't go after them.[152]

Scherer says he enjoys selling so much that he is sorry he didn't get into that aspect of the business earlier. (He had been in engineering, manufacturing and management for twenty years before becoming a salesman.)

I feel I'm a fairly good salesman. I sell with honesty. I've got a reputation in the industry of being honest and knowing a little bit about it. If I don't know...[I'll] say I don't and try to get the information.... However, if I have to lie to the customer I don't want the order. Because I don't sell for today, I sell for tomorrow.[153]

One of the most interesting stories about the industry Scherer says occurred during his first stint with Westcote. Scherer also sold a sheet product for a company in Chicago called Weldon Industries owned by John McDougal:

[And] he sold [his company] to, let's say some product company, a well-known company. He [McDougal] couldn't go back in business [for] five or seven years. [Then] he went back in business and maybe four or five years later [the same company] bought him out again![154]

[1]Norman manuscript, p. 5.

[2]*Ibid.*, p. 9.

[3]Gray interview, p. 18.

[4]*Ibid.*

[5]Jim Smith interview, p. 14.

[6]*Ibid.*, p. 8.

[7]*Ibid.*

[8]Carlson interview, p. 30.

[9]Gray interview, p. 14.

[10]*Ibid.*, p. 15.

[11]Wert interview, p. 28, 29.

[12]*Ibid.*, p. 32.

[13]Eismann interview, p. 21.

[14]Karl and Erik Zimmer interview with Bill Klein, p. 12.

[15]*Ibid.*, p. 27.

[16]*Ibid.*, p. 10.

[17]*Ibid.*, p. 9.

[18]Beck interview, p.11.

[19]Skip Heintzelman interview with Bill Klein, P. 11.

[20]*Ibid.*, p. 12.

[21]*Ibid.*, p. 7.

[22]*Ibid.*, p. 3, 4.

[23]*Ibid.*, p. 12.

[24]*Ibid.*, p. 12, 13.

[25]*Ibid.*, p. 25, 26.

[26]*Ibid.*, p. 21.

[27]*Ibid.*, p. 27.

[28]Sobel interview, p. 17, 18.

[29]Schwartz interview, p. 26.

[30]Dochstader interview, p. 11, 12.

[31]*Ibid.*, p. 14.

[32]*Ibid.*, p. 15, 18.

[33]*Ibid.*, p. 19.

[34]*Ibid.*, p. 20.

[35]English interview, p. 28, 29.

[36]Bunnell interview, p. 19.

[37]Prittie interview, p. 19. Both cation (positive) and anion (negative) are UV inks.

[38]*Ibid.*, p. 20.

[39]Deckard interview, p. 14.

[40]Gray interview, p. 20.

[41]Gartner interview, p. 16.

[42]*Ibid.*, p. 18.

[43]Gray interview, p. 2.

[44]*Ibid.*, p. 11, 12.

[45]*Ibid.*, p. 21.

[46]*Ibid.*, p. 19.

[47]Rüesch interview, p. 22, 23.

[48]*Ibid.*, p. 23.

[49]*Ibid.*, p. 26.

[50]Hart interview, p. 14, 15.

[51]*Ibid.*, p. 16.

[52]*Ibid.*, p. 13.

[53]Bob Sirois interview with Bill Klein, p. 2.

[54]*Ibid.*, p. 2.

[55]Barry McKillip interview with Bill Klein, p. 13.

[56]*Ibid.*, p. 15.

[57]*Ibid.*, p. 17.

[58]Sirois interveiw, p. 12.

[59]*Ibid.*, p. 13, 15.

[60]*Ibid.*, p. 16.

[61]Andrews, Jr. interview, p. 31.

[62]*Ibid.*, p. 28.

[63]Bunnel interview, p. 15, 16.

[64]Prittie interview, p. 23.

[65]Dochstader interview, p. 22.

[66]Garber interview, p. 5.

[67]*Ibid.*, p. 8, 9.

[68]*Ibid.*, p. 10.

[69]*Ibid.*, p. 12.

[70]*Ibid.*, p. 6. A company in England came out with a linerless label concept in about 1989.

[71]Wes Nedblake interview with Bill Klein, p. 19.

[72]*Ibid.*, p. 7.

[73]*Ibid.*, p. 8.

[74]*Ibid.*

[75]*Ibid.*, p. 9, 10.

[76]*Ibid.*, p. 20.

[77]*Ibid.*, p. 5.

[78]*Ibid.*, p. 27.

[79]George Reinke interview with Bill Klein, p. 2.

[80]*Ibid.*

[81]*Ibid.*, p. 3.

[82]*Ibid.*, p. 3, 4.

[83]*Ibid.*, p. 8.

[84]Campbell interview, p. 13, 14.

[85]*Ibid.*, p. 16.

[86]*Ibid.*, p. 20.

[87]*Ibid.*, p. 23.

[88]*Ibid.*, p. 24.

[89]*Ibid.*, p. 25, 26.

[90]Jules Farkas interview with Bill Klein, p. 6.

[91]*Ibid.*

[92]*Ibid.*, p. 9, 10.

[93]*Ibid.*, p. ll.

[94]Reed interview, p. 23.

[95]Bill Klein in Wise interview, p. 19, 20.

[96]Capuzzo interview, p. 18, p. 19.

[97]*Ibid.*, p. 25.

[98]*Ibid.*

[99]Bill Klein in Jim Wilson interview, p. 8.

[100]Jim Hattemer correspondence with Bill Klein, p. 4, 5.

[101]Yeagle interview, p. 14.

[102]Jim Wilson interview, p. 8.

[103]Gartner interview, p. 14, 17.

[104]Hart interview, p. 9.

[105]Gartner interview, p. 15.

[106]Werneke interview, p. 16.

[107]*Ibid.*, p. 24.

[108]*Ibid.*, p. 26.

[109]Gray interview, p. 26.

[110]Bunnell interview, p. 18.

[111]Tomlinson interview, p. 22.

[112]Bill Klein in Desanzo interview, p. 18.

[113]Wise interview, p. 20.

[114] Clark, First Fifty Years, p. 116.

[115]David Good interview with Bill Klein, p. 5.

[116]*Ibid.*, p. 6.

[117]Craig Daniell interview with Bill Klein, p. 2.

[118]*Ibid,* p. 3.

[119]*Ibid.*, p. 9.

[120]*Ibid.*, p. 21.

[121]*Ibid.*, p. 26, 22.

[122]Carlson interview, p. 23

[123]Bunnell interview, p. 12.

[124]*Ibid.*, p. 13

[125]Prittie interview, p. 17.

[126]Good interview, p. 8.

[127]*Ibid.*, p. 4, 5, 8.

[128]*Ibid.*, p. 9.

[129]Don McDaniel quote in Bill Whitfield interview with Bill Klein, p. 10.

[130]Whitfield interview, p. 2.

[131]*Ibid.*

[132]*Ibid.*, p. 2, 3.

[133]*Ibid.*, p. 4.

[134]*Ibid.*, p. 5.

[135]*Ibid.*, p. 14.

[136]*Ibid.*, p. 15.

[137]*Ibid.*, p. 18, 19

[138]"CDM Quotes" received by Bill Klein.

[139]*Ibid.*

[140]*Ibid.*

[141]Rick Spear interview with Bill Klein, p. 4.

[142]*Ibid.*, p. 4, 5.

[143]*Ibid.*, p. 10.

[144]*Ibid.*, p. 12.

[145]*Ibid.*, p. 15.

[146]*Ibid.*, p. 20, 21, 23.

[147]*Ibid.*, p. 27.

[148]*Ibid.*, p. 43, 44.

[149]*Ibid.*, p. 44.

[150]This figure includes other label-like configurations such as transdermals, security codes, etc.

[151]Scherer interview, p. 7.

[152]*Ibid.*, p. 19.

[153]*Ibid.*, p. 17, 20, 21.

[154]*Ibid.*, p. 8.